LIBRARY
ZION-BENTON TOWNSHIP
ZION, ILLINOIS

W9-AZV-893

TWENTIETH CENTURY INTERPRETATIONS
OF

THE FALL
of the
HOUSE
OF USHER

A Collection of Critical Essays

Edited by
THOMAS WOODSON

Prentice-Hall, Inc. A SPECTRUM BOOK *Englewood Cliffs, N. J.*

LIBRARY
ZION-BENTON TOWNSHIP HIGH SCHOOL
ZION, ILLINOIS

813
P74w

For Nancy

Copyright © 1969 by Prentice-Hall, Inc., Englewood Cliffs, New Jersey. A SPEC-TRUM BOOK. All rights reserved. No part of this book may be reproduced in any form or by any means without permission in writing from the publisher. P 13-301721-4; C 13-301739-7. *Library of Congress Catalog Card Number 69-15343.* Printed in the United States of America.

Current printing (last number):
10   9   8   7   6   5   4   3   2   1

Prentice-Hall International, Inc. (*London*)

Contents

iii

Introduction

by Thomas Woodson

I

Edgar Allan Poe is still, almost a century and a quarter after his sudden and mysterious death in Baltimore in 1849, the most controversial of American writers. At first, it was Poe's life that became a literary spectacle. To his "official" biographer, Rufus Griswold, Poe was at times the incarnation of evil, more dangerous than the common run of criminals because of his literary abilities. But to his early and influential admirer, the great French poet Charles Baudelaire, Poe was one of the most profoundly human men of letters of the nineteenth century, precisely because of his aura of vision and vice: "I prefer Edgar Poe, drunk, poor, persecuted, and a pariah, to a calm and virtuous Goethe or Walter Scott. I should willingly say of him and of a special class of men what the catechism says of our Lord: 'He has suffered much for us.' "[1] The public could make its choice between Satan and Christ.

Later biographies present Poe as a more normal human being, and one scholar has gone so far as to describe his life as one of the dullest of his time.[2] But it is likely that his readers will never fully accept such a judgement because the nature of Poe's genius was to suggest, through his stories and poems, an exciting world unlike any that his readers—particularly American readers—would ever live in. Nevertheless, this remote world is paradoxically familiar (is perhaps the most familiar form of art to a popular audience) because of detective stories and the horror movies revived so often on late evening television, where it develops a theme increasingly compelling in our time: the

[1] Lois and Francis E. Hyslop, Jr., trans. and eds., *Baudelaire on Poe: Critical Papers* (State College, Penn.: Bald Eagle Press, 1952), p. 85.

[2] Edward H. Davidson, "Introduction," *Selected Writings of Edgar Allan Poe* (Boston: Houghton Mifflin Company, 1956), p. viii.

1

theme of apocalypse, the ending of things, the collapse of all constructions at the moment of a final atomic explosion.

In the twentieth century the focal point of the Poe controversy has moved from his life to his work. The story of melodramatic destruction that Poe's contemporaries accepted as a limited, but legitimate, kind of literature, seemed to be—particularly after the fall of millions of houses and families in the First World War—the work of a dangerous adolescent playing around with matters of life and death. Henry James and T. S. Eliot argued that Poe's work was just what a mature literary mind should not take seriously. In his freewheeling discussion of earlier American writers in *The Green Hills of Africa,* Ernest Hemingway put it quite bluntly: "Poe is a skilful writer. It [his writing] is skilful, marvelously constructed, and it is dead." Though few would argue that writing about death must necessarily itself be dead, modern expectations about fiction turned away from Poe, as the comments of Brooks, Warren, Gordon, and Tate (reprinted here) demonstrate. Poe's exaggerated mannerisms of style and plot seemed unreal, no longer expressing any humanity.

But the fascination with Poe, first felt abroad by Baudelaire, continued and gained power. Mallarmé, in an elegant phrase, claimed that Poe's style had purified, renewed and civilized the language of literature, instead of killing it. D. H. Lawrence, an acute commentator on the modern world and the American literary tradition, anticipated and anatomized the critical attacks on Poe by the New Critics: "Moralists have always wondered helplessly why Poe's 'morbid' tales need have been written. They need to be written because old things need to die and disintegrate, because the old white psyche has to be gradually broken down before anything else can come to pass. . . . For the human soul must suffer its own disintegration, *consciously,* if ever it is to survive." Although Lawrence explicitly denies to Poe the Christ-like renewing power of art, the rhythm of "forming of a new consciousness underneath" the old consciousness it rejects, he nevertheless finally sees him as an "adventurer into vaults and cellars and horrible underground passages of the human soul," not as the cold-minded Satan who systematically reduces the soul to nothing. Poe's concern for literature as originality, as an adventure toward renewal of soul, is, Charles Feidelson argues, what unites him with his American contemporaries Emerson and Whitman, however eccentric his practice, for his stories present "not simply the emotion of horror but the irrational state of mind, terrified at itself, yet oddly prolific." The mere fact that the controversy continues, and that Poe's defenders are holding their own (or perhaps more) indicates that the stories have life, are an aesthetic reality that is enduring and perhaps even central to American experience.

"The Fall of the House of Usher" presents Poe's art at its best. Although it was not his favorite story (this was "Ligeia"), and such influential readers as Washington Irving and Allen Tate have preferred "William Wilson," it combines the psychological qualities of these stories with the more immediate popularity of "The Raven" and "The Gold Bug." The physical collapse of the decayed mansion into the lake beneath has a universal archetypal appeal, as I have already hinted. No work of Poe's has so haunted later artists and readers. It led Debussy to attempt the musical setting of "a symphony on psychologically developed themes," and later an opera, which obsessed him for years, but was never accomplished. The story has been perhaps even more provocative visually: Jean Epstein's silent film in 1928 contributed to the history of cinematic innovation. There is also a Hollywood version, that of Roger Corman in 1960. In a more subtle way the house of Usher has kept a place among the images every artist and poet retains below the level of his consciousness; Louis Rubin has suggested that the "voices singing out of empty cisterns and exhausted wells" of T. S. Eliot's "Wasteland" derive ultimately from Poe's description of the house.[3] Paul Ramsey has argued that the paintings of Roderick Usher constitute "Poe's strange invention of abstract art," "an uncanny illumination, and foreshadowing" of the nonobjective painting of Picasso and Mondrian, and of Pollock and Kandinsky. Usher's (and Poe's) purpose, Ramsey claims, agrees with this statement by the Expressionist Franz Marc: "One is no longer concerned with the reproduction of Nature, but destroys it in order to show the mighty laws that surge from behind the beautiful appearances of things." More than any other single work of Poe's, "Usher" anticipates the violent freedom of modern art.[4]

Poe apparently wrote "Usher" in the spring or summer of 1839 in Philadelphia. He published it in September of that year in *Burton's Gentleman's Magazine,* of which he was assistant editor. He included

[3] Louis Rubin, *The Curious Death of the Novel* (Baton Rouge: Louisiana University Press, 1967), p. 58.
[4] Paul Ramsey, Jr., "Poe and Modern Art," *College Art Journal,* XVIII (1959), 210–15.

it in a collection of twenty-five stories, *Tales of the Grotesque and Arabesque,* published in December by Lea and Blanchard, also of Philadelphia. The story had nothing overtly in common with his life at the time, which was quietly domestic. He lived in a cottage with his very young wife, Virginia, and her mother, Mrs. Maria Clemm, in the outskirts of the city. He walked to and from his office downtown daily. William Burton, the English actor who employed him, had contracted with Poe in July 1839 to "edit" his new magazine, which meant handling the soliciting and organizing of articles and reviews, proofreading, overlooking the actual printing, and—incidentally—providing original "articles" of fiction and review from his own pen. For all this Poe received a salary of $50 per month.

Such shabby treatment was not new to Poe; in fact, he was rarely paid decently for anything he wrote; but his menial duties did not impede his own creativity: each month from August to December he printed a story in *Burton's*: "The Man That Was Used Up," a satirical burlesque; "The Fall of the House of Usher"; "William Wilson"; "Morella" (reprinted from an earlier publication); and "The Conversation of Eiros and Charmion," a visionary philosophical dialogue. These, along with the rest of his short fiction to date, all went into the *Tales,* which he hoped would make his reputation as a great artist in prose. But the publishers were willing to bring out the book only on the condition that the author have no share of the profits from sales. They were prudent; there was little profit.

Poe was thirty years old when he wrote "Usher." Though one of his biographers has described his work during 1839 as "the first golden milestone" of his career,[5] it is more realistic to see it as one of a succession of futile efforts to establish himself, to define a solid identity for posterity, for his contemporaries, and for himself. It is of course typical of him that he should try to construct his own literary personality by dramatizing the fall of a house and of a family.

II

Edgar Poe was born in 1809 in Boston, where his parents, Elizabeth Arnold Poe and David Poe, were impoverished actors. David Poe was an undistinguished performer, tending to alcoholism, proud that his

---

[5] James A. Harrison, "Biography," *The Complete Works of Edgar Allan Poe,* ed. James A. Harrison (New York: Crowell Collier and Macmillan, Inc. 1902), I, 148.

father "General" Poe had attained distinction during the Revolution as a purveyer of supplies to Lafayette and other commanders. What roots the family had were in the South, in Maryland. Before Edgar was a year old, his father disappeared while he and his wife were playing in New York, and was never heard of again. Mrs. Poe took her children to Richmond, Virginia, where she sickened and died of tuberculosis in December 1811. Edgar was taken in by a neighbor couple, Mr. and Mrs. John Allan. Allan, a Scottish-born tobacco and cotton merchant, an aggressive self-made businessman, became his guardian and the dominant male influence of his formative years, though he never formally adopted him. Edgar Poe became Edgar Allan Poe at this time. Mrs. Allan, a passive, affectionate, and child-less woman, became his substitute mother.

John Allan's solid bourgeois temperament was bound to conflict with that of a sensitive, emotional, and impractical boy. In 1815 the Allans took him to England, where they lived for five years, and Poe attended grammar school in Scotland and the public school at Stoke Newington, near London (one of the few events of his life that he used directly in his fiction, in "William Wilson"). Poe was a good student, especially in languages, but he was spoiled by his foster father's generous allowance; he sensed himself different from his schoolmates, anticipating the trials of identity which were to come.

When Poe entered the University of Virginia in 1826 at the age of seventeen, his serious troubles with Allan began. In some ways their relationship suffered from a "generation gap," though Poe did not intend to "drop out," nor did he at this time turn to drugs, as his stories might suggest. Despite the legend that he was expelled for continual "debauchery," the fact is that Poe left the university after one term because Allan refused to provide him sufficient funds for necessary expenses, thus leading him into foolish gambling. After his pleas for help were unanswered, he left Richmond, working his way by ship to Boston under an assumed name. There he arranged to publish anonymously his first book of poems, and enlisted in the U.S. Army, again under a false name. He was soon assigned to Fort Moultrie, South Carolina, where Hervey Allen thinks he may have seen the prototype of the house of Usher—"some old, crumbling, and cracked-walled mansion . . . surrounded by its swamps and gloomy woods, its cypress-stained tarns, and its snake-haunted Indian moats." [6]

---

[6] Hervey Allen, *Israfel: The Life and Times of Edgar Allan Poe* (New York: Rinehart and Company, 1927), I, 220.

Mrs. Allan died in 1829, at about the time that Poe was discharged honorably from the army. He applied for admission to West Point, hoping a serious military career would bring a reconciliation with John Allan. Allan had originally intended to bring him into his business, and Poe's distaste for trade continued to annoy him. Poe's second small volume of poems came out in Baltimore in 1829, when he went to live there with his aunt, Mrs. Maria Clemm, and her six-year-old daughter, Virginia, while awaiting his appointment to West Point.

His stay at West Point lasted less than a year. Again, Allan's indifference led him to give up his military career; he deliberately disobeyed orders, and was dismissed in March 1831. With a characteristic flourish, he published this same month in New York his *Poems, Second Edition,* dedicated to the U.S. Corps of Cadets. He returned to Mrs. Clemm's house in Baltimore.

At this time his career as a writer of stories began. In 1831, there was, as now, considerably more public attention paid to fiction than to poetry; Washington Irving had become the first American writer of international reputation through his *Sketch-Book* (1819); Nathaniel Hawthorne had turned from an abortive start as a novelist to the short story, and his first efforts were just then appearing in New England newspapers and magazines. In fact, the vogue of the story, particularly in the horrific or Gothic mode, was moving from Britain, where it was well established (for example, in *Blackwood's Edinburgh Magazine*), to the receptive ranks of new American periodicals.

Poe's first stories were written in 1831 for a contest sponsored by a Philadelphia newspaper. It is significant that he began not as an imitator of the Gothic fad, but by satirizing and burlesquing it. Though he did not win the prize, the stories were published. In 1833, he did win a similar prize, sponsored by a Baltimore paper with his "Ms. Found in a Bottle," the visionary narrative of a fantastic voyage to destruction that remains one of his most interesting works. John Pendleton Kennedy, a Baltimore lawyer and part-time novelist, was on the prize committee; he recognized Poe's genius, and introduced him to Thomas W. White, a printer, and owner of *The Southern Literary Messenger,* a new magazine in Richmond. In a short time Poe rose to become its editor (though White wanted to reserve that title for himself); his career as a professional man of letters was now genuinely begun. In less than two years on the *Messenger* he gained a national reputation, principally as a reviewer and critic, increasing the magazine's circulation from 500 to 3,500, making it the leading critical journal in the South. As a critic, he was incisive, harsh, and sometimes insulting, but his intelligence and learning gave most of his reviews high quality and made them clearly the best literary journalism that had yet appeared in the United States.

In spite of this success Poe never won White's admiration; his salary remained close to $10 a week. White disliked the literary controversies he stirred up, and Poe responded to this by beginning to drink heavily and to miss deadlines. White was a man of John Allan's mold (Allan had died in 1834 without leaving Poe a cent of his considerable fortune), a small-town merchant with little appreciation of literature. In January 1837 Poe resigned his editorship and moved to New York, but literary work proved hard to find there, in part because of the economic depression of that year. He did however manage to complete his novel *The Narrative of Arthur Gordon Pym,* an uneven but remarkably imaginative sea-story that anticipates some aspects of Melville's *Moby-Dick.* Shortly after it was published in July 1838, he moved to Philadelphia, where he was to write "Ligeia" that summer, and "The Fall of the House of Usher" the following year.

Poe's relationship with women is one of the most often debated sides of his life and personality. Characters like Ligeia and Madeline Usher have led critics, particularly those of a Freudian persuasion, to see him as a classic case of Oedipal fixation. Following the trauma of his mother's death, he spent his childhood with Frances Allan. In his early teens he seems to have felt an intense admiration for Mrs. Jane Stith Stanard of Richmond, the mother of one of his classmates. She died suddenly when he was fifteen. By the time he entered the University at seventeen, he had become engaged to Sarah Elmira Royster, a girl of about his age. But when Allan refused to pay his student debts, Sarah's parents broke off the engagement.

None of these relationships was abnormal, considering that Poe grew up an orphan. But his marriage in 1835 to Virginia Clemm, his thirteen-year-old cousin, has caused endless speculation about his alleged tendency to incest, or, perhaps, his impotence. Marie Bonaparte, in her psychoanalytical study of Poe, finds the love of Edgar and Virginia essentially incestuous, like that of the "identical" twins Roderick and Madeline Usher. But the fact is that an immediate reason for their marriage was Poe's attempt to rescue the impoverished

Virginia and her mother from having to go to live with his philistine cousin Neilson Poe. He brought them from Baltimore to Richmond at the time when he was becoming established with the *Messenger*.

Edgar and Virginia Poe had no children, and it is entirely possible that the marriage was never consummated. Virginia was a sickly girl, stricken with a tubercular hemorrhage in 1842, and seriously ill from then until her death in 1847. But Poe's idealization of her, and the insubstantial quality of his literary heroines, is only slightly out of the ordinary for his time, the early years of the Victorian era. We can see a comparable attitude in Hawthorne's letters to his wife, and in his romantic heroines Phoebe, Priscilla, and Hilda.

In 1840 Poe broke with Burton, as he had with White, and sought to begin his own *Penn Magazine,* but he could not obtain sufficient financial support to start publication. Frustrated of the income he needed to write with leisure, he applied without success for work at the Philadelphia custom-house (Hawthorne held such a position in Boston, and later in Salem). In 1841 and 1842 he edited *Graham's Magazine,* one of the most distinguished of American periodicals, but after building its circulation, he again quarreled with the owner about editorial policy, and again unsuccessfully tried to start a periodical of his own, to be called *The Stylus.* He continued to write stories, among them "Murders in the Rue Morgue" and "The Gold Bug," but it was not until he had moved his family to New York in 1844 that his poem "The Raven" finally brought him fame. In 1845 he published a new volume of poems, featuring "The Raven," and a volume of *Tales,* in which "Usher" was revised to its final form.

Thanks now to the assistance of James Russell Lowell, Poe found employment on *The Broadway Journal,* a literary weekly, and finally attained his dream of ownership in October 1845. By the following January, however, he had to stop publication because of illness and failing funds. Virginia's disease, his business disappointments and perhaps the romantic fever of living too intensely in the imagination led him to heavier drinking; after Virginia died, early in 1847, he published less. He seemed to feel a sense of coming death, which he somewhat restlessly countered by becoming romantically involved successively with three women, one of them his boyhood sweetheart Sarah Royster, now a widow. The principal literary accomplishment of his last years was *Eureka,* a pseudoscientific treatise on cosmology. It is also a remarkable prose poem, and has been used increasingly in recent years to explain the inner imaginative life of his stories.

Poe was travelling from Richmond, where he intended to marry Sarah Elmira Royster Shelton, to New York, where Mrs. Clemm was waiting to return with him, when he was found unconscious in a street in Baltimore on October 3, 1849. Taken to a hospital, he regained consciousness in a state of "busy but not violent delirium— constant talking—and vacant converse with spectral and imaginary objects on the walls." [7] A prolonged delirium followed, and he died on October 7 without regaining full consciousness. There has never been a conclusive explanation of what caused his collapse and death. The legend of course mentions alcohol and drugs. It seems likely that liquor affected him much more immediately than most men, but it does not seem to have dominated his life. Nevertheless, the images that afflicted him on his deathbed inevitably recall the unforgettable fantasies of his fiction.

III

Before he committed "The Fall of the House of Usher" to *Burton's Magazine,* Poe sent it to *The Southern Literary Messenger,* hoping that T. W. White might print it there. His quarrel with White of two years before had cooled somewhat, and the *Messenger* continued to enjoy the reputation and circulation Poe had brought to it. But neither White nor his editorial assistant, James Heath, was interested in "Usher"; Heath's reply is an interesting comment on the contemporary literary market:

[White] is apprehensive . . . that the "Fall of the House of Usher" would not only occupy more space than he can conveniently spare (the demands upon his columns being very great), but that the subject-matter is not such as would be acceptable to a large majority of his readers. He doubts whether the readers of the "Messenger" have much relish for tales of the German school, although written with great power and ability. . . . I doubt very much whether tales of the wild, improbable, and terrible class can ever be permanently popular in this country. Charles Dickens it appears to me has given the final death-blow to writings of that description. . . .[8]

[7] Letter of Dr. J. J. Moran to Mrs. Maria Clemm, November 15, 1849; Poe, *Complete Works,* I, 335.
[8] Letter of Heath to Poe, September 12, 1839; Poe, *Complete Works,* XVII, 48.

Heath's prediction had a point. The pendulum of popular taste was moving away from the story of supernatural atmosphere, and Dickens was leading the way. His *Pickwick Papers*, a novel illustrative of every-day life and everyday people, humorous, full of vivid and precise facts reported with exact observation, had sold 40,000 magazines a month in its serialized form in 1836. *Oliver Twist*, which added a sharp and detailed attack on the Poor Law to realistic setting and characteriza-tion, also did well in serial from 1837 to 1839. Lack of international copyright law allowed American publishers to pirate such English fiction with impunity.

Heath's letter was in effect an answer to a letter Poe had sent to White four years earlier, just before he joined the *Messenger*. Defend-ing his horrifying story "Berenice," he had claimed:

> The history of all Magazines shows plainly that those which have at-tained celebrity were indebted for it to articles *similar in nature—to Berenice*—although, I grant you, far superior in style and execution. I say similar in *nature*. You ask me in what does this nature exist? In the ludicrous heightened into the grotesque: the fearful coloured into the horrible: the witty exaggerated into the burlesque: the singular wrought out into the strange and mystical.[9]

He goes on to cite several titles in *Blackwood's* and *The London Magazine*, among them "The Confessions of an English Opium Eater," by Thomas DeQuincey (at that time attributed to the great Samuel Taylor Coleridge).

Poe's comment had a point, too. The tale of terror had enjoyed a good deal of influence in the history of popular journalism, and was able to present psychological truths that Dickens' social realism could not touch. Unfortunately, it already seemed old-fashioned, for it was now seventy years since Horace Walpole had started the mode with his *Castle of Otranto: A Gothic Story* (in 1765). "Gothic," though at first an architectural term, referring to the soaring churches of the Middle Ages, considered wild and barbarous by Renaissance critics,— had come in the mid-eighteenth century to describe an aspect of the reaction against rationalism. Devotees of Gothic gloried in the con-templation of the medievally remote and mysterious—in ruins,

---

[9] Letter of Poe to Thomas W. White, April 30, 1835; John Ward Ostrom, ed., *The Letters of Edgar Allan Poe* (Cambridge, Mass.: Harvard University Press, 1948), I, 57–58.

tombs, and ghosts; in the light of the moon dimly seen amid the crypts and corridors of crumbling palaces; in mysterious haunted strangers and beautiful women stalked by enigmatic figures with un-natural or supernatural lusts and powers. The most successful English writer in the Gothic vein, Mrs. Anne Radcliffe, completed her work before 1800. During the first two decades of the nineteenth century several German Romantic writers had also exploited the possibilities of terror, though frequently by substituting materials from local folk-lore for the sensational paraphernalia of the English. The "German school" to which Heath refers in his letter included E. T. A. Hoffmann, Tieck, la Motte Fouqué, and Achim von Arnim. Excerpts from their fiction had appeared in English in the 1820's, especially in Thomas Carlyle's collection, *German Romance* (1827).

By the 1830's elements of burlesque, hoax, and parody had entered the genre. Poe began his career in fiction by exploiting this side of Gothic, though for him (and others) its conventions were ambivalent, and could be pushed either toward the ludicrous or towards greater seriousness. In America, particularly, Poe could see that few writers had made really imaginative use of its possibilities. Washington Irving, and even Charles Brockden Brown (whose novels dated from the last years of the previous century), had treated Gothic subjects, but with a heavier, more moralistic touch than Poe thought artistically proper. And Irving's understanding of the darker side of human psychology remained superficial.

Poe had written a number of stories before "Usher"—"Metzenger-stein," "The Assignation," "Berenice," "Morella," "Ligeia"—in which he had elicited effects of horror from Gothic settings. The geography of these settings is generally imprecise, or only precise enough to harmon-ize with the reader's expectations of the exotic, the "German," and the Gothic: "Ligeia," for instance, is first set in "the lonely desolation" of a dwelling in a "dim and decaying city by the Rhine," and then moves to "the gloomy and dreary grandeur" of an ancient abbey "in one of the wildest and least frequented portions of fair England." What is important is the isolation that allows Poe to remove all social questions and focus entirely on the mental dramas of his protagonists. The locale of the House of Usher is completely unspecified; only the story's emphasis on the antiquity of the building and the family leads the reader to suspect an English or German setting.

We can see Poe's originality in handling conventional materials by

comparing "The Fall of the House of Usher" with the German story
that most resembles it in setting and that some scholars have argued
is its source: E. T. A. Hoffmann's "The Entail" ("Das Majorat,"
1817).[10] "The Entail" is the story of the ancestral castle of a noble
family, located on the shore of the Baltic Sea in a desolate environment, reflecting the morose and antisocial temperament of a baron
named Roderick, who forces his descendants to keep the castle by
converting it into a property of entail (a property given to a succession
of heirs, which they may not sell or give away). The narrator, a romantic young man whose uncle is legal adviser to the family, accompanies
his uncle on a visit to the castle; shortly after, while reading a ghost
story at night (compare the function of the "Mad Trist" of Sir Launcelot Canning in "Usher"), he hears an inhuman shriek from behind the
walls, apparently the voice of an evil spirit trying to escape. This turns
out—much later—to be the ghost of an old servant who had murdered
Baron Roderick's son for failing to accept the entail. The narrator
meanwhile falls sentimentally in love with the lady of the house, the
Baroness Seraphina, a beautiful but nervous girl who suffers from
extreme morbid excitability, and who eventually falls victim to the
dark destiny of the entail—she suddenly stands screaming in a moving sledge and is thrown to the ground dead. There are suggestions of
Roderick and Madeline Usher in these characters, but the method of
narration and the personality of the narrator are quite different from
Poe's. Hoffman was a lawyer as well as a romantic storyteller; he exploits the effects of entail law and the development of strife within
the family in a quite realistic way; further, the narrator's uncle, the
judge, is a warm and ironically witty commentator on this young
man's rather silly pretentions. Thus the tone is complex, and the story's
telling takes about five times as much space as "Usher."

There is reason to suspect that "The Entail" had something to do
with Poe's intentions. It had been summarized and quoted at length
by Sir Walter Scott in an important article on Hoffmann, "On the
Supernatural in Fictitious Composition" (1827), which Poe very probably read carefully.[11] Scott translated the opening description of the

[10] The connection was first noticed by E. C. Stedman in his edition of Poe's
Works (Chicago: Stone and Kimball, 1894), I, 97. See also Henry A. Pochmann,
German Culture in America (Madison: University of Wisconsin Press, 1957), pp.
403, 719. Pochmann argues that the *plot* of "Usher" is closer to that of a story by
Achim von Arnim, "The Lords of Entail" ("Die Majoratsherren," 1820).

[11] Gustav Gruener, "E. T. A. Hoffmann and Edgar Allan Poe," *PMLA*, XIX
(1904), 16, and Pochmann, *German Culture,* pp. 403, 715, 719.

castle, emphasizing its psychological power more than the original context had: "Part of the castle was in ruins; a tower built for the purpose of astrology by one of its old possessors, the founder of the majorat in question, had fallen down, and by its fall made a deep chasm, which extended from the highest turret down to the dungeon of the castle." Scott did not entirely approve of Hoffmann, finding his work less a model for imitation than "as affording a warning how the most fertile fancy may be exhausted by the lavish prodigality of its possessor." Poe was not likely to be so cautious, but he was indebted to Scott's review nevertheless. Scott commented of Hoffmann: "The grotesque in his composition resembles the arabesque in painting, in which is introduced the strange and complicated monsters . . . and all other creatures of romantic imagination, dazzling the beholder as it were by the unbounded fertility of the author's imagination." Scott called Hoffmann "the inventor [of] the fantastic or supernatural grotesque."[12] This term "grotesque" pleased Poe much more than "German" or "Gothic"; he called his first book of stories, including "Usher," *Tales of the Grotesque and Arabesque,* probably following Scott.

In his preface to this book Poe makes some important statements about his art. But he does not exactly define the terms *grotesque* and *arabesque*. He begins: "The epithets 'Grotesque' and 'Arabesque' will be found to indicate with sufficient precision the prevalent tenor of the tales here published," and goes on to say, "I am led to think it is this prevalence of the 'Arabesque' in my serious tales, which has induced one or two critics to tax me, in all friendliness, with what they have been pleased to term 'Germanism' and gloom." This last statement, plus the phrase, "the ludicrous heightened into the grotesque" in the letter to White I have quoted, have led Arthur Hobson Quinn to the distinction that ". . . generally speaking the Arabesques are the product of powerful imagination and the Grotesques have a burlesque or satirical quality."[13] Most of Poe's critics have accepted this idea, classifying "Usher" as an example of arabesque, perhaps particularly because Poe's narrator says of Roderick Usher's flowing silken

[12] Sir Walter Scott, "The Novels of E. T. A. Hoffmann," *Miscellaneous Prose Works* (Edinburgh, 1835), XVII, 311, 332, 306–7. Originally published as "On the Supernatural in Fictitious Composition . . . ," *Foreign Quarterly Review,* I (July 1827), 60–98.

[13] Arthur Hobson Quinn, *Edgar Allan Poe* (New York: Appleton-Century-Crofts, 1941), p. 289.

hair: ". . . I could not, even with effort, connect its Arabesque expression with any idea of simple humanity."

The term *arabesque* comes ultimately from Arabian and Moorish art, and generally refers to a kind of fluid, wildly complicated design in which human and animal figures are sometimes introduced among intertwined branches, foliage, and fanciful scrollwork. *Grotesque,* however, is a more ambiguous term. Wolfgang Kayser, who has studied the history of this concept in art and literature, disagrees with Quinn's distinction, finding that Poe, following Scott's discussion of Hoffmann, "uses grotesque and arabesque synonymously in the title of his collection." [14] Kayser argues that Poe's handling of the grotesque is more than comic and satirical; in fact, Kayser's description of how the artists of the Renaissance looked at this mode seems remarkably relevant to the inner meaning of "The Fall of the House of Usher": the grotesque is

> a specific ornamental style suggested by antiquity, . . . not only something playfully gay and carelessly fantastic, but also something ominous and sinister, in the face of a world totally different from the familiar one—a world in which the realm of inanimate things is no longer separated from those of plants, animals, and human beings, and where the laws of statics, symmetry, and proportion are no longer valid . . . the sphere in which the dissolution of reality and the participation in a different kind of existence, as illustrated by the ornamental grotesques, form an experience about the nature and significance of which mankind has never ceased to ponder.[15]

The theme of "sentience" in Poe's story, the strange sense of life in the house and its surroundings, is thus part of a serious and traditional vision of reality, rather than simply a trumped up and meretricious "atmosphere," as Poe's detractors have usually claimed.

Poe certainly felt that his stories were serious and realistic in the highest sense. He concluded his preface with an eloquent defense against the charge of Germanism and pseudohorror: "If in many of

[14] Wolfgang Kayser, *The Grotesque in Art and Literature,* trans. Ulrich Weisstein (Bloomington: Indiana University Press, 1963), p. 77. Kayser quotes a passage from "The Masque of the Red Death" in which Poe again uses the terms synonymously.
[15] Kayser, *The Grotesque,* pp. 21–22. Writers of the Romantic period, and particularly Poe, used the idea of the grotesque to describe the most serious of their writings, according to Arthur Clayborough, *The Grotesque in English Literature* (London: Oxford University Press, 1965), p. 11.

my productions terror has been the thesis, I maintain that terror is not of Germany, but of the soul,—that I have deduced this terror only from its legitimate sources, and urged it only to its legitimate results." His final answer to the cavils of White and Heath, to the social realism of Dickens, and to the demanding moralism of American critics from James Russell Lowell to R. P. Blackmur, is here.

IV

One of Poe's most characteristic ideas is unity. This is the key to the philosophy of *Eureka* and the visionary dialogues; this is the essence of his contribution to the theory and practice of the short story. In his review of Hawthorne's *Twice-Told Tales* he speaks of the need for totality, for "a certain unique single *effect* to be wrought out": every word of a piece of fiction should contribute to the realization of this aim. Many critics have noticed the intensity of the concentration on unity in "Usher"—how the opening description of the House and the narrator's reactions lead swiftly and inevitably to the final catastrophe. It is perhaps instructive to see the story's structure as three sections of about equal length: paragraphs 1–14 introduce the House, the circumstances of the narrator's visit, and Roderick and Madeline Usher; the movement is of entrance, of the opening of doors (there is perhaps a pun on the family name when the valet "threw open a door and ushered me into the presence of his master"). The middle section, paragraphs 15 to 23, develops the narrator's initial impressions through the analysis of Roderick's aesthetic ideas, and the application of his most abstract and intellectual madness to his relationship with his sister. The final part, paragraphs 24 to 40, logically completes these themes by building rapidly to Madeline's reappearance and the simultaneous collapse of both Ushers and their house. Here there is a final door-opening, as Madeline reels on the threshold; then the tarn finally "closes" over the fragments of the fallen house. As often in Poe's fiction, the protagonist (here the narrator and both Ushers combined) ends teetering on the verge of a supreme revelation that is also his destruction, an opening that mockingly also closes everything.

Poe attains an effect of structural symmetry by placing certain actions and images only in the first and third parts: the sullen tarn; the appearances of Madeline in the apartment; and the speeches of Roderick (there is little dialogue; Roderick speaks only twice at length, once early in the story to confirm his "intolerable agitation of soul," and then at the end to announce, *"We have put her living in the tomb!"*).

This unity of structure is reinforced by the style. From the magnificent opening sentences to the end, Poe's language is highly charged with emotion, but at the same time, quite abstract: for instance, the narrator remarks at the beginning, "a sense of insufferable gloom pervaded my spirit." Instead of simple, direct description, we experience the narrator's *sense* of things. *Pervaded* is also a key word, evoking the notorious "atmosphere" of the House, the "condensation" of "vapours" from the tarn. As Poe's note about "sentience" suggests, he seeks credibility for his psychological claims by using the vocabulary of nineteenth-century chemistry. *Pervaded* is also important in introducing the theme of *oppression,* which recurs throughout: something sinister presses down on the mind, anticipating the collapse of the House. Later the narrator says: "An irrepressible terror pervaded my frame; and, at length, there sat upon my very heart an incubus of utterly causeless alarm." *Frame* hints at the identity of the human body with the building, as does the similarity of appearance between Roderick's head and the House, and the imagery of his poem, "The Haunted Palace." In fact, a close study of the story's style reveals a very high degree of recurrence of a rather small and special vocabulary. The effect of this stylistic narrowness is to bring together, almost to the verge of solipsism, the sensations of both Roderick and the narrator, and the "sentience" of Madeline and the House, Madeline being less a character than an object to be perceived.

Another peculiarity of Poe's diction is his emphasis on the negative. In the first paragraph alone we find, in addition to *insufferable,* these adjectives and verbs: *unrelieved, unredeemed, unnerved, insoluble,* and *unsatisfactory*. The story is full of words beginning with *un-* or *in-,* or ending in *-less,* and expressions containing *no, not, mere,* or *scarcely*. Poe uses these words to maintain an excited, exaggerated tone, but also to evoke the results of oppression on the mind: the nightmare of a vacant, featureless world—imaged by sinking beneath the surface of the tarn—a world where meaning and value have dissolved into nothingness.

Through these devices Poe has made style his weapon to redeem, as William Hedges has put it, the sensational material of popular fiction for the analysis of the soul.[16] In the speeches of Roderick Usher this style is speeded up, intensified, almost to the point of hysterical incoherence. This style is then counterpointed to the burlesque extravagance of the "Mad Trist" of Sir Launcelot Canning, which the narrator describes as a style of "uncouth and unimaginative prolixity," portentously empty of meaning (perhaps Poe's attempt to imitate the effect of the porter's knocking on the door in *Macbeth*).

We can see the extent to which Poe was conscious of style if we compare his practice in "Usher" to the comment on and burlesque of the conventional Gothic tale that he wrote only a year before. In "How to Write a Blackwood Article," his farcical narrator, Signora Psyche Zenobia (alias Suky Snobbs), is advised by "Mr. Blackwood" that the secret of "your writer of intensities" is to give a "record of his sensations. Sensations are the great thing after all. Should you ever be drowned or hung, be sure to take note of your sensations—they will be worth to you ten guineas a sheet." He goes on to counsel the proper styles for a successful story. Among these is "the tone elevated, diffusive, and interjectional": "Some of our best novelists patronize this tone. The words must be all in a whirl, like a humming-top, and make a noise very similar, which answers remarkably well instead of meaning."

Attending to her lesson, Psyche Zenobia produces a model story, "A Predicament (or the Scythe of Time)." The first paragraph of this forms an interesting comparison with the first paragraph of "Usher." (And its plot is a close parody of "The Pit and the Pendulum," which Poe did not write until five years later!) The restless, baffled, but probing syntax of the horseman arriving at the House of Usher is, in effect, parodied by this "record of sensations." But self-parody is an index to self-awareness, at times, and the passages differ considerably in prose rhythm.[17] Poe is an accomplished craftsman at maintaining a level of style consonant with his intention and theme.

[16] William L. Hedges, *Washington Irving* (Baltimore: Johns Hopkins Press, 1965), p. 203.
[17] See Donald Barlow Stauffer, "Style and Meaning in 'Ligeia' and 'William Wilson,'" *Studies in Short Fiction*, II (1965), 316–30, for an account of how Poe varies rhythm in burlesque and serious stories.

V

Although Roderick Usher is certainly one of the most memorable characters in American literature, and a forerunner of the modernist heroes of James and Proust, Poe is not primarily interested in characterization. Whereas Dickens in the *Pickwick Papers* presents a cast seemingly of thousands, each with his eccentric trait or "humor," Poe's drive for brevity and concentration in style leads to a curious situation in which the *identity* of mind and object, of actor and sufferer, becomes a passion. Dickens' fiction tends towards the conditions of theater, but Poe's tends toward the conditions of lyric poetry. Despite the narrator's frequent attempts at dispassionate analysis of the problem, from the outset the style identifies his sensations with those of Roderick. To use Poe's own words (about Roderick and Madeline), "sympathies of a scarcely intelligible nature exist between them." In a sense, the description of Roderick's *physique* in paragraph 8 is no more than an extension of what the narrator has said of the exterior and interior of the House. Madeline, moreover, is not mentioned until paragraph 13, simultaneous with her "appearance," which is physically in "a remote portion of the apartment" just as psychologically she exists at this time in a remote portion of the narrator's experience. Roderick, we have noticed, says little, and the narrator's point of view leads us to define him more through his choice of reading and music, his poem, and his paintings, than by more natural methods of characterization.

Thus Poe presents really only one character, of which Roderick, Madeline, the narrator—and the House—are different aspects. The narrator feels "a sense of *insufferable* gloom" at the outset; Roderick has become a prisoner of the House's influence "by dint of long *sufferance*." But Roderick is not only passive before this oppressive power; he is part of it: he is like the tarn in that from his mind "darkness, as if an inherent positive quality, poured forth upon all objects of the moral and physical universe in one unceasing radiation of gloom." In Poe's cosmology life is a continual process of attraction and repulsion, of radiation and absorption (of someone or thing being "pervaded" by someone or thing), and such is the case in this story as well.

The names of such abstract, almost allegorical characters may be

especially significant. Thomas O. Mabbott has found that Poe knew a
family named Usher during his childhood; they were actors, friends
of his mother, who cared for her during her last illness. Their children, a twin brother and sister of about Poe's age, were of neurotic
personality; both died young.[18] This may be the source of the name;
I have suggested earlier that Poe may be using the etymology of
"usher"; it seems also possible that he had in the back of his mind the
medieval ballad, "The Wife of Usher's Well," in which the ghosts
of three drowned sons return to visit their mother.

Roderick, Mabbott also claims, is named for "the last of the Visi-
goths," a character well known to contemporary readers from a poem
by the minor British writer Alaric A. Watts. Roderick is likewise the
name of the last lord of the castle in Hoffmann's "The Entail." "Madeline seems to be named in a most indirect way for St. Mary of Magdala.
Magdala is a tower; she is 'lady of a house.' " [19] This name may also
recall ironically the heroine of Keats' "Eve of St. Agnes," who escapes
the castle that imprisons her because of the power of her innocent
love. In any case, the Ushers are determinedly "literary" in origin
and being, more teasing to the mind, perhaps, because of their insub-
stantiality and abstractness.

VI

Poe has held a precarious position in the canon of "classical"
American writers, because of the extremity to which he pushed certain
tendencies in Romantic literary art. Vernon Parrington, for example,
found no place for him in his chronicle of the liberal spirit in American literature. But in a real sense Poe's role is that of explorer, voy-
ager, pioneer. We can see more clearly because of the intensity of
"The Fall of the House of Usher" how the struggle with nothingness
experienced by the heroes of Hawthorne's and Melville's fiction is
central to American experience. Roderick Usher's obsessions illuminate
the problems of Arthur Dimmesdale in *The Scarlet Letter,* Clifford
Pyncheon in *The House of the Seven Gables,* and Captain Ahab in
Moby-Dick. The House of Usher as an archetypal symbol prefigures

[18] Thomas O. Mabbott, "The Fall of the House of Usher," *The Explicator,* XV
(1956), No. 7.
[19] Thomas O. Mabbott, "Poe's Vaults," *Notes and Queries,* CXCVIII (1953),
542–43.

the house of Pyncheon in Hawthorne, Gilbert Osmond's Roman villa in James' *Portrait of a Lady,* and the slowly declining house of Compson in Faulkner's *The Sound and the Fury.* We may even see the House as comparable to the whaleship *Pequod* in *Moby-Dick,* that cannibal of a craft, obsessed with its mission of destruction until it too is pulled down by the encompassing ocean.

Beyond American perspectives, Poe's story has a place among the most compelling images of international Romanticism. For instance, Floyd Stovall has compared the "sentience" of the House to that of Wordsworth's "Ruined Cottage," [20] and Leo Spitzer, in the essay reprinted here, has found striking stylistic parallels with the Pension Vauquer in Balzac's *Père Goriot.*

To a considerable extent, twentieth-century interpretations of "The Fall of the House of Usher" reveal the moral crisis that Romantic literature has forced upon our literary criticism. Those who find Poe's art false are taking their cue, it appears to me, from the reaction to Romanticism voiced most strongly in the 1920's by Irving Babbitt. In his essay, "Coleridge and Imagination" (1929), Babbitt attacked what he considered the decivilizing surrender to primitive emotion in much of English Romantic poetry, and particularly in Coleridge's "Rime of the Ancient Mariner." For the Mariner, like the multiple protagonist of "Usher,"

> does not do anything. In the literal sense of the word he is not an agent, but a patient. The true protagonists of *The Ancient Mariner,* Professor Lowes remarks rightly, are the elements—"Earth, Air, Fire, and Water in the multiform balefulness and beauty." . . . Between a poem, like *The Ancient Mariner,* in which the unifying element is feeling and a poem which has a true unity of action the difference is one of kind; between it, and let us say, *The Fall of the House of Usher* the difference is at most one of degree. In this and other tales Poe has, like Coleridge, and indeed partly under his influence, achieved a unity of tone or impression, a technique in short, perfectly suited to the shift of the center of interest from action to emotion.[21]

We might answer Babbitt that for both Coleridge and Poe there is meaningful action in the hero's submission to and participation in

[20] Floyd Stovall, "The Conscious Art of Edgar Allan Poe," in *Poe: A Collection of Critical Essays,* ed. Robert Regan (Englewood Cliffs, N.J.: Prentice-Hall, Inc., 1967), p. 176.

[21] Irving Babbitt, "Coleridge and Imagination," *The Bookman,* LXX (1929), 113–24.

the elemental powers. It is the powerful honesty of Poe's vision of life that led him to show us a mocking nightmare as the "legitimate results" of imagination, rather than a hollow triumph of the will and an easy, shallow paradise, such as we find too frequently in the general run of American and modern literature.

PART ONE

View Points

Arthur Hobson Quinn

In September [1839], Poe presented *Burton's* with one of his very greatest stories, "The Fall of the House of Usher." It continues the treatment of the theme of identity, but this time the fear of Roderick Usher that the building will decay deepens his terror of the loss of identity into his apprehension of racial ruin. At the very outset of the story the atmosphere of tragedy is established with consummate art. Through the eyes of the narrator, that nameless person who is so much more real than he is usually credited with being, the house becomes alive with meaning. The very bareness and desolation are active forces calling up those unusual emotions which only in the hands of a master can spring from the contemplation of ordinary things. The effort of the visitor to rearrange his view of these forbidding objects and thus destroy this effect of desolation is shown to be fruitless. One of the most common errors in Poe criticism lies in the assumption of the absence of heart in his characters. But the narrator has come a long distance simply because the appeal of Roderick Usher has clutched at his friendship through that quality—"It was the apparent *heart* that went with his request." Roderick and he are bound with the tie that is next to love and family affection, the friendship that comes only in early youth, before distrust becomes a duty.

The relation between Roderick and Madeline, his twin sister, is once more an identity of a strange and baffling kind. Her disease threatens that identity; but her death restores it in another world. Poe's own knowledge of drawing creates an intriguing episode in the painting of the long tunnel, lighted by rays that admit of no explanation. Every sense, therefore, sight and hearing especially, are keyed above the normal, and lead to the poetic fantasy which had earlier been published as "The Haunted Palace," but which is here ascribed

From Edgar Allan Poe: A Critical Biography *by Arthur Hobson Quinn (New York: Appleton-Century-Crofts, 1941), pp. 284–85. Copyright 1941 by D. Appleton-Century Co., Inc. Reprinted by permission of Appleton-Century-Crofts.*

to Roderick and intensifies his mood of terror. Poe transferred to Roderick his own fear of impending mental decay which came at times during his life. The loss of spiritual identity is naturally the final human danger, and Roderick lives in its shadow. The approach to the climax of the story, through the entombment of Madeline, her rising from the coffin and her return to her brother just before her real death, is controlled by Poe with a skill of which his earlier stories of premature burial had given promise. The madness of Roderick, the flight of the visitor, and the rending of the house into ruin are portrayed with that economy of which Poe alone among writers of the short story was at that time possessed.

Cleanth Brooks and Robert Penn Warren

This is a story of horror, and the author has used nearly every kind of device at his disposal in order to stimulate a sense of horror in the reader: not only is the action itself horrible but the descriptions of the decayed house, the gloomy landscape in which it is located, the furnishings of its shadowy interior, the ghastly and unnatural storm— all of these are used to build up in the reader the sense of something mysterious and unnatural. Within its limits, the story is rather successful in inducing in the reader the sense of nightmare; that is, if the reader allows himself to enter into the mood of the story, the mood infects him rather successfully.

But one usually does not find nightmares pleasant, and though there is an element of horror in many of the great works of literature— Dante's *Inferno*, Shakespeare's tragedies—still, we do not value the sense of horror for its own sake. What is the meaning, the justification of the horror in this story? Does the story have a meaning, or is the horror essentially meaningless, horror aroused for its own sake?

In the beginning of the story, the narrator says of the House of Usher that he experienced "a sense of insufferable gloom," a feeling which had nothing of that "half-pleasurable, because poetic, sentiment with which the mind usually receives even the sternest natural images of the desolate or terrible. . . . It was a mystery all insoluble. . . ."

From Understanding Fiction *by Cleanth Brooks and Robert Penn Warren* (New York: Appleton-Century-Crofts, 1943), pp. 202–5. Copyright 1943 by F. S. Crofts and Co., Inc. Reprinted by permission of Appleton-Century-Crofts.

Does the reader feel, with regard to the story as a whole, what the narrator in the story feels toward the house in the story?

One point to determine is the quality of the horror—whether it is merely vague and nameless, or an effect of a much more precise and special imaginative perception. Here, the description which fills the story will be helpful: the horror apparently springs from a perception of decay, a decay which constitutes a kind of life-in-death, monstrous because it represents death and yet pulsates with a special vitality of its own. For example, the house itself gets its peculiar *atmosphere* . . . from its ability apparently to defy reality: to remain intact and yet seem to be completely decayed in every detail. By the same token, Roderick Usher has a wild vitality, a preternatural acuteness and sensitiveness which itself springs from the fact that he is sick unto death. Indeed, Roderick Usher is more than once in the story compared to the house, and by more subtle hints, by implications of descriptive detail, throughout the story, the house is identified with its heir and owner. For example, the house is twice described as having "vacant eye-like windows"—the house, it is suggested, is like a man. Or, again, the mad song, which Roderick Usher sings with evident reference to himself, describes a man under the *allegory* . . . of a house. To repeat, the action of the story, the description, and the symbolism, consistently insist upon the horror as that which springs from the unnatural and monstrous. One might reasonably conclude that the "meaning" of the story lies in its perception of the dangers of divorcement from reality and the attempt to live in an unreal world of the past, or in any private and abstract world of thought. Certainly, elements of such a critique are to be found in the story. But their mere presence there does not in itself justify the pertinacious and almost loving care with which Poe conjures up for us the sense of the horrors of the dying House of Usher.

One may penetrate perhaps further into the question by considering the relation of the horror to Roderick Usher himself. The story is obviously his story. It is not Madeline's—in the story she hardly exists for us as a human being—nor is it the narrator's story, though his relation to the occupants of the doomed House of Usher becomes most important when we attempt to judge the ultimate success or failure of the story.

Roderick Usher, it is important to notice, recognizes the morbidity of the life which he is leading. Indeed, he even calls his persistence in carrying on his mode of living in the house a deplorable "folly." And yet one has little or no sense in the story that Roderick Usher is actually making any attempt to get away from the haunting and oppressive gloom of the place. Actually, there is abundant evidence that he is in love with "the morbid acuteness of the senses" which he has cultivated in the gloomy mansion, and that in choosing between this and the honest daylight of the outside world, there is but one choice for him. But, in stating what might be called the moral issue of the story in such terms as these, we have perhaps already overstated it. The reader gets no sense of struggle, no sense of real choice at all. Rather, Roderick Usher impresses the reader as being as thoroughly doomed as the decaying house in which he lives.

One may go further with this point: we hardly take Roderick Usher seriously as a real human being at all. Even on the part of the narrator who tells us that Usher has been one of his intimate companions in boyhood, there is little imaginative identification of his interests and feelings with those of Usher. At the beginning of the story, the narrator admits that he "really knew little" of his friend. Even his interest in Usher's character tends to be what may be called a "clinical" interest. Now, baffled by the vague terrors and superstitions that beset Roderick Usher, he is able to furnish us, not so much a reading of his friend's character as a list of symptoms and aberrations. Usher is a medical case, a fascinating case to be sure, a titillatingly horrible case, but merely another case after all.

In making these points, we are really raising questions that have to do with the limits of tragedy. The tragic protagonist must be a man who engages our own interests and hopes and fears as a Macbeth or a Lear, of superhuman stature though these be, engages them. We must not merely look on from without. Even if the interest is an overwhelming psychological interest, it is not enough. Eugene O'Neill perhaps falls into the same error in handling his protagonist in *The Emperor Jones*. We see the overlayings of civilization progressively torn from the big confident Negro, until at the end he is reduced to an impotent wreck, at the mercy of the primitive superstitions which he himself had thought he had put away. But the play is hardly a tragedy; for there is no imaginative identification of ourselves with the protagonist; there is merely an outside interest, clinical observation.

In the case of Roderick Usher, then, there is on our part little imaginative sympathy and there is, on his own part, very little struggle. The story lacks tragic quality. One can go farther: the story lacks even pathos—that is, a feeling of pity, as for the misfortune of a weak person, or the death of a child. To sum up, Poe has narrowed the fate of his protagonist from a universal thing into something special and even peculiar, and he has played up the sense of gloom and monstrous derangement so heavily that free will and rational decision hardly exist in the nightmare world which he describes. The horror is relatively meaningless—it is generated for its own sake; and one is inclined to feel that Poe's own interest in the story was a morbid interest.

Marie Bonaparte

The mansion, Lady Madeline's double in so far as she herself is a mother-symbol to Poe-Usher, also repeats her fate in this sudden dissolution. The narrator-friend, Usher's double, escapes from death—or rather from the dead and avenging mother, who has seized the latter from beyond the grave—since someone had to be left to tell the story.

But the deeper meaning of this sinister tale lies in the fate of Usher. Poe is *punished* for having betrayed his mother in loving Madeline-Virginia. Usher-Poe is *punished* for not having dared to seek and rewin his babyhood mother when, like Annabel Lee, men bore her away, and also for his silence and acceptance, in his childish incomprehension of death. Usher-Poe is *punished* for his sadism, as shown in the way Usher treated his sister. Finally, Usher-Poe is *punished* for his infantile incestuous wishes towards his mother, as witness all the quotations from the *Mad Trist*. The legendary theme of the dragon, which must be killed to win some woman—with or without the aid of treasure—is as old as the world. It is a perfect expression of the Oedipus wish: the dragon, symbol of the father, is killed and the mother set free to belong to the victorious son. It is the theme of the legend of Perseus and Andromeda, of Siegfried and Brunhild. The woman in the extracts quoted from the *Mad Trist* remains hidden, but this is due to the strength of Poe's sexual repressions. That Ethelred should force his way into the dwelling of the hermit—but another father-figure—by an act which may also be taken as the symbol of a sexual attack upon

From The Life and Works of Edgar Allan Poe: a Psycho-Analytic Interpretation *by Marie Bonaparte, trans. John Rodker (London: Imago Publishing Co., Ltd., 1949), pp. 249–50. Originally published in French in 1933. Copyright 1949 by Imago Publishing Co. Ltd. Reprinted by permission of Hillary House Publishers Ltd., Hogarth Press Ltd., and the Literary Estate of the author.*

the mother,[1] that he should slay the fire-breathing dragon and possess himself of the magic shield, can only be motivated by an ulterior aim —the conquest of a woman which, being prohibited, brings its own punishment therewith.

When the Lady Madeline, representative, as it were, of the deadly mansion, returns from the tomb to seek her brother, it is as the emissary of justice. Nevertheless, Poe's phantasy of the mother who will return from the grave to find her son and claim him in death—a phantasy which was to dog his unconscious to the day, when in Baltimore, it came to pass—was not only a phantasy of retribution, but one of wish-fulfilment. All neurotic symptoms and phantasies, however, similarly develop from two conflicting factors. The horrible death of which Madeline is the instrument, constitutes her brother's punishment in this life, though at the same time making possible the gratification of his desires in that other "life-in-death" which will thenceforth be his. There, at last, he might sing, paraphrasing the ending of *Annabel Lee*—

> And so, all the night-tide, I lie down by the side
> Of my darling—my darling—my life and my bride,
> In her sepulchre there *in the tarn,*
> *In the depths of the stagnant tarn*—

That tarn, in which the House of Usher sleeps forever: brother with sister and mother with son.

Caroline Gordon and Allen Tate

This famous story is perhaps not Poe's best, but for the purposes of this book it has significant features which ought to illuminate some of the later, more mature work in the naturalistic-symbolic technique of Flaubert, Joyce, and James. Poe's insistence upon the unity of effect, from first word to last, in the famous review of Hawthorne . . . anticipates from one point of view the high claims of James in his

---

Reprinted with the permission of Charles Scribner's Sons from The House of Fiction *(New York: Charles Scribner's Sons, 1950), pp. 114–16, by Caroline Gordon and Allen Tate. Copyright 1950 Charles Scribner's Sons.*

[1] Cf. Freud, *Analysis of a Phobia in a Five-year-old Boy, Collected Papers,* III, p. 149. London: The Hogarth Press and the Institute of Psycho-Analysis. Translated from *Analyse der Phobie eines fünfjährigen Knaben,* 1909, *Ges. Werke,* Band VII.

essay "The Art of Fiction." James asserts that the imaginative writer must take his art at least as seriously as the historian takes his; that is to say, he must no longer apologize, he must not say "it *may* have happened this way"; he must, since he cannot rely upon the reader's acceptance of known historical incident, create the illusion of reality, so that the reader may have a "direct impression" of it. It was towards this complete achievement of "direct impression" that Poe was moving, in his tales and in his criticism; he, like Hawthorne, was a great forerunner. The reasons why he did not himself fully achieve it (perhaps less even than Hawthorne) are perceptible in "The Fall of the House of Usher."

Like Hawthorne again, Poe seems to have been very little influenced by the common-sense realism of the eighteenth-century English novel. What has been known in our time as the romantic sensibility reached him from two directions: the Gothic tale of Walpole and Monk Lewis, and the poetry of Coleridge. Roderick Usher is a "Gothic" character taken seriously; that is to say, Poe takes the Gothic setting, with all its machinery and *décor,* and the preposterous Gothic hero, and transforms them into the material of serious literary art. Usher becomes the prototype of the Joycean and Jamesian hero who cannot function in the ordinary world. He has two characteristic traits of this later fictional hero of our own time. First, he is afflicted with the split personality of the manic depressive:

> His action was alternately vivacious and sullen. His voice varied rapidly from a tremulous indecision (when the animal spirits seemed utterly in abeyance) to that species of energetic concision . . . and perfectly modulated guttural utterance, which may be observed in the lost drunkard, or the irreclaimable eater of opium, during the periods of his most intense excitement.

Secondly, certain musical sounds (for some unmusical reason Poe selects the notes of the guitar) are alone tolerable to him: "He suffered from a morbid acuteness of the senses." He cannot thus live in the real world; he is constantly exacerbated. At the same time he "has a passionate devotion to the intricacies . . . of musical science"; and his paintings are "pure abstractions" which have "an intensity of intolerable awe."

Usher is, of course, both our old and our new friend; his new name is Monsieur Teste, and much of the history of modern French literature is in that name. Usher's "want of moral energy," along with a hypertrophy of sensibility and intellect in a split personality, places him in the ancestry of Gabriel Conroy, John Marcher, J. Alfred Prufrock, Mrs. Dalloway—a forbear of whose somewhat tawdry accessories they might well be a little ashamed; or they might enjoy a degree of moral complacency in contemplating their own luck in having had greater literary artists than Poe present them to us in a more credible imaginative reality.

We have referred to the Gothic trappings and the poetry of Coleridge as the sources of Poe's romanticism. In trying to understand the kind of unity of effect that Poe demanded of the writer of fiction we must bear in mind two things. First, unity of *plot,* the emphasis upon which led him to the invention of the "tale of ratiocination"; but plot is not so necessary to the serious story of moral perversion of which "The Fall of the House of Usher," "Ligeia," and "Morella" are the supreme examples. Secondly, unity of tone . . . , a quality that had not been consciously aimed at in fiction before Poe. It is this particular kind of unity, a poetical rather than a fictional characteristic, which Poe must have got from the Romantic poets, Coleridge especially, and from Coleridge's criticism as well as "Kubla Khan" and "Christabel." Unity of plot and tone can exist without the *created, active detail* which came into this tradition of fiction with Flaubert, to be perfected later by James, Chekhov, and Joyce.

For example, in "The Fall of the House of Usher," *there is not one instance of dramatized detail.* Although Poe's first person narrator is in direct contact with the scene, he merely reports it; he does not show us scene and character in action; it is all description. The closest approach in the entire story to active detail is the glimpse, at the beginning, that the narrator gives us of the furtive doctor as he passes him on "one of the staircases." If we contrast the remoteness of Poe's reporting in the entire range of this story with the brilliant recreation of the character of Michael Furey by Gretta Conroy in "The Dead," we shall be able to form some conception of the advance in the techniques of reality that was achieved in the seventy-odd years between Poe and Joyce. The powerful description of the façade of the House of Usher, as the narrator approaches it, sets up unity of tone, but the description is never woven into the action of the story: the "metaphysical" identity of scene and character reaches our consciousness through *lyrical assertion.* The fissure in the wall of the house remains an inert symbol of Usher's split personality. At the climax of the story Poe uses an incredibly clumsy device in the effort to make the collapse of Usher active dramatically; that is, he employs the mechanical device of coincidence. The narrator is reading to Usher the absurd tale of the "Mad Trist" of Sir Launcelot Canning. The knight has slain the dragon and now approaches the "brazen shield," which falls with tremendous clatter. Usher has been "hearing" it, but what he has been actually hearing is the rending of the lid of his sister Madeline's coffin and the grating of the iron door of the tomb; until at the end the sister (who has been in a cataleptic trance) stands outside Usher's door. The door opens; she stands before them. The narrator flees and the House of Usher, collapsing, sinks forever with its master into the waters of the "tarn."

We could dwell upon the symbolism . . . of the identity of house and master, of the burial alive of Madeline, of the fissure in the wall of the house and the fissure in the psyche of Usher. What we should emphasize here is the dominance of symbolism over its visible base: symbolism external and "lyrical," not intrinsic and dramatic. The active structure of the story is mechanical and thus negligible; but its lyrical structure is impressive. Poe's plots seem most successful when the reality of scene and character is of secondary importance in the total effect; that is, in the tale of "ratiocination." He seemed unable to combine incident with his gift for "insight symbolism"; as a result his symbolic tales are insecurely based upon scenic reality. But the insight was great. In Roderick Usher, as we have said, we get for the first time the archetypal hero of modern fiction. In the history of literature the discoverer of the subject is almost never the perfecter of the techniques for making the subject real.

Harry Levin

Poe's dramatism tends toward monodrama, in which the sole actor is acted upon by chimeras, like Flaubert's Saint Anthony. Faced with such isolation, Hawthorne's egoists had been urged to forget themselves in the idea of another, in love and marriage and domesticity. But that idea was foredoomed, with Poe, to throw him back upon himself.

*   *   *

The most persistent motif of his prose, as of his poetry, is what might—all too etymologically—be termed the posthumous heroine.

*   *   *

The pattern is most spectacularly embodied in "The Fall of the House of Usher," where it has been visualized through the impressionable sensibilities of a spectator, and relayed to us "with an utter depression of soul which [he] can compare to no earthly sensation more properly than to the after-dream of the reveller upon opium." Poe is therewith afforded his best opportunity for an atmospheric presentation, in which scenic detail is artfully confounded with emotional reaction. The house of Usher is, ambiguously, "both the family and the family mansion," the stately gloom of the building itself presaging the decadence of its inhabitants. The pallid and cadaverous Roderick Usher and his twin sister, the Lady Madeline, happen to be the last of their ancient and inbred race. "Sympathies of a scarcely intelligible nature had always existed between them," we are told, with Byronic innuendo. Roderick, languidly poring over his esoteric books, is a Hamlet whose artistic gifts have been introverted by "the grim phantasm Fear." The most describable of his abstract paintings brightly depicts an inaccessible tunnel. Of the dirges he wildly improvises, to the accompaniment of his rhapsodic guitar, Poe reprints one, the poem he elsewhere printed as "The Haunted Palace." By a similar transposition, "The Conqueror Worm" is reintroduced as a poetic commentary upon "Ligeia." Another poem, "Israfel," is linked with "The Fall of the House of Usher" through the quoted metaphor comparing the heart to a lute. Roderick's heartstrings vibrate to the decline, the catalepsy, and the interment of Madeline. Heralded by a reading from a romance, she emerges from the vault in her bloody shroud, and brother and sister share their death-agonies.

Shakespeare's image for the gates of the sepulcher, "ponderous and marble jaws," is significantly modified by Poe, who replaces "marble" with "ebony." The symbolism of his interpolated stanzas—eclipsing the downright melodrama of Thomas Hood's stanzas on "The Haunted House"—rests upon the analogy between the façade of a palace and the face of a man, between the house and the brain. In Poe's theoretical phraseology, Roderick's *morale* is influenced by the *physique* of his dwelling, from which for years he has not ventured forth. Its "vacant eye-like windows," which strike the approaching observer, foreshadow the mental condition of its occupant. The final collapse of its ruins into the tarn, while the retreating narrator looks

From The Power of Blackness *by Harry Levin (New York: Alfred A. Knopf, Inc., 1958), pp. 154, 156, 159–61.* © *Copyright 1958 by Harry Levin. Reprinted by permission of Harry Levin and Alfred A. Knopf, Inc.*

back through the moonlit storm at the desolated landscape, projects
an apocalypse of the mind. So much is explicit in the verse and prose
of the tale; more is implicit if we look ahead, or if we relocate Poe's
Gothic terrors within a regional perspective. Much that seems forced,
in William Faulkner's work, becomes second nature when we think
of him as Poe's inheritor. We think of Caddy and Quentin, those two
doomed siblings of the house of Compson, or of Emily Grierson, that
old maid who clings to the corpse of her lover. In retrospect, Poe's
work acquires a sociological meaning when it is linked with the culture of the plantation in its feudal pride and its foreboding of doom.
But there is still another sense in which Roderick Usher with all his
idiosyncrasies, awaiting his own death and hastening that of his sister,
prefigures a larger and nearer situation: the accomplished heir of all
the ages, the hypersensitive end-product of civilization itself, driven
underground by the pressure of fear.

Wayne C. Booth

So far we have considered only commentary which is about something clearly dramatized in the work. The authors have simply tried
to make clear to us the nature of the dramatic object itself, by giving us the hard facts, by establishing a world of norms, by relating
particulars to those norms, or by relating the story to general truths.
In so doing, authors are in effect exercising careful control over the
reader's degree of involvement in or distance from the events of
the story, by insuring that the reader views the materials with the
degree of detachment or sympathy felt by the implied author.

A different element enters when an author intrudes to address the
reader's moods and emotions directly. "There are certain themes of
which the interest is all-absorbing, but which are too entirely horrible
for the purposes of legitimate fiction. . . . To be buried while alive
is, beyond question, the most terrific of the extremes which has ever
fallen to the lot of mere mortality." When Poe begins "The Premature
Burial" (1844) in this way and continues for several pages with talk
about the horror of premature burial, and about its frequency, we

Reprinted from The Rhetoric of Fiction *by Wayne C. Booth (Chicago: The
University of Chicago Press, 1961), pp. 200–203, by permission of The University
of Chicago Press.* © *1961 by The University of Chicago.*

feel that something is wrong. He is addressing us directly, immediately, attempting to put us into a frame of mind *before* his story begins; it is difficult for us to resist boredom or annoyance. "Fearful indeed the suspicion [that such events occur]—but more fearful the doom! It may be asserted, without hesitation, that *no* event is so terribly well adapted to inspire the supremeness of bodily and of mental distress, as is burial before death." Whatever the effect of this kind of thing on Poe's original magazine readers, one can hardly believe that experienced readers have ever been very strongly moved by it.

We might at first be tempted to blame the superlatives; after all, one remembers so many other "supreme" horrors in other Poe stories. But such superlatives would be much more acceptable if reserved to describe the actual plight of the victim during his interment. Just as Melville's "Shakespearean" commentary seems ludicrously exaggerated when read in extracts from *Moby Dick* but usually seems unobjectionable and appropriate in context, so this prose, bad as it seems in isolation, might in a proper setting be acceptable. But the story provides it with no context. It is isolated rhetoric, the author in his own name and person doing what he can, with all the stops pulled, to work us into a proper mood before his story begins. "Get ready to shudder," he seems to say, and like the voice of the commentator in a bad documentary film, he is divorced from the effects of his own rhetoric.

If we compare this with the fully integrated mood-building of a better story, "The Fall of the House of Usher" (1839), we see one reason for the frequent insistence that indispensable commentary be spoken by a character in the story.

> During the whole of a dull, dark, and soundless day in the autumn of the year, when the clouds hung oppressively low in the heavens, I had been passing alone, on horseback, through a singularly dreary tract of country, and at length found myself, as the shades of the evening drew on, within view of the melancholy House of Usher. I know not how it was—but, with the first glimpse of the building, a sense of insufferable gloom pervaded my spirit. I say insufferable; for the feeling was unrelieved by any of that half-pleasurable, because poetic, sentiment, with which the mind usually receives even the sternest natural images of the desolate or terrible.

By the simple expedient of creating a character who experiences the rhetoric in his own person, it has been made less objectionable. Every adjective and detail intended to set our mood is a part of the growing mood and experience of the central character; the rhetoric now seems functional, "intrinsic." It is no longer simply directed outward—as if it were a drug that could be injected into the spectator on his way into the theater.

We might easily make the mistake, however, of generalizing falsely from this comparison. It does not follow either that commentary is always effective so long as it is spoken by a character in the story or that this story would be further improved by revealing more and more of its tone through dramatized detail and less and less through narrative statement. Caroline Gordon and Allen Tate view this story as an important step, but a step only, in the grand progress toward the mastery of *"creative, active detail* which came into this tradition of fiction with Flaubert, to be perfected later by James, Chekhov, and Joyce."[1] To them, since the story has "not one instance of dramatized detail," it is still only half-realized. What they are really asking is that *all* general commentary, unrelieved by irony, should be eliminated. The narrator must not say "bleak walls," or "vacant eye-like windows," or "black and lurid tarn that lay in unruffled lustre." The walls and windows and tarn should be dramatically portrayed in order to be made visually alive with their bleakness and vacuity and luridness *shown* to the reader rather than merely *told*. This seems to me a demand that springs from the prejudices of an age desiring effects basically different from Poe's. For Poe's special kind of morbid horror, a psychological detail, as conveyed by an emotionally charged adjective, is more effective than mere sensual description in any form. Whatever may be wrong with "The Fall of the House of Usher" is not to be cured by changing the technique. If I am now unable to react as Poe intended, it seems quite clear that I would not do so no matter what technique he used. Those of us who can remember a time when Poe *was* effective know how indispensable the heavy adjectives are.

[1] *The House of Fiction* (New York, 1950), p. 116.

PART TWO

Interpretations

Edgar Allan Poe

by D. H. Lawrence

Poe has no truck with Indians or Nature. He makes no bones about Red Brothers and Wigwams.

He is absolutely concerned with the disintegration-processes of his own psyche. As we have said, the rhythm of American art-activity is dual.

1. A disintegrating and sloughing of the old consciousness.
2. The forming of a new consciousness underneath.

Fenimore Cooper has the two vibrations going on together. Poe has only one, only the disintegrative vibration. This makes him almost more a scientist than an artist.

Moralists have always wondered helplessly why Poe's "morbid" tales need have been written. They need to be written because old things need to die and disintegrate, because the old white psyche has to be gradually broken down before anything else can come to pass.

Man must be stripped even of himself. And it is a painful, sometimes a ghastly process.

Poe had a pretty bitter doom. Doomed to seethe down his soul in a great continuous convulsion of disintegration, and doomed to register the process. And then doomed to be abused for it, when he had performed some of the bitterest tasks of human experience, that can be asked of a man. Necessary tasks, too. For the human soul must suffer its own disintegration, *consciously,* if ever it is to survive.

But Poe is rather a scientist than an artist. He is reducing his own self as a scientist reduces a salt in a crucible. It is an almost chemical

From Studies in Classic American Literature *by D. H. Lawrence (New York: Thomas Seltzer, Inc., 1922). Copyright 1923 by Thomas Seltzer, Inc. Copyright 1951 by Frieda Lawrence. Reprinted by permission of The Viking Press, Inc. and Laurence Pollinger Ltd.*

analysis of the soul and consciousness. Whereas in true art there is always the double rhythm of creating and destroying.

This is why Poe calls his things "tales." They are a concatenation of cause and effect.

His best pieces, however, are not tales. They are more. They are ghastly stories of the human soul in its disruptive throes.

Moreover, they are "love" stories.

Ligeia and *The Fall of the House of Usher* are really love stories.

Love is the mysterious vital attraction which draws things together, closer, closer together. For this reason sex is the actual crisis of love. For in sex the two blood-systems, in the male and female, concentrate and come into contact, the merest film intervening. Yet if the intervening film breaks down, it is death.

So there you are. There is a limit to everything. There is a limit to love.

The central law of all organic life is that each organism is intrinsically isolate and single in itself.

The moment its isolation breaks down, and there comes an actual mixing and confusion, death sets in.

This is true of every individual organism, from man to amoeba.

But the secondary law of all organic life is that each organism only lives through contact with other matter, assimilation, and contact with other life, which means assimilation of new vibrations, non-material. Each individual organism is vivified by intimate contact with fellow organisms: up to a certain point.

So man. He breathes the air into him, he swallows food and water. But more than this. He takes into him the life of his fellow men, with whom he comes into contact, and he gives back life to them. This contact draws nearer and nearer, as the intimacy increases. When it is a whole contact, we call it love. Men live by food, but die if they eat too much. Men live by love, but die, or cause death, if they love too much.

There are two loves: sacred and profane, spiritual and sensual.

In sensual love, it is the two blood-systems, the man's and the woman's, which sweep up into pure contact, and *almost* fuse. Almost mingle. Never quite. There is always the finest imaginable wall between the two blood-waves, through which pass unknown vibrations, forces, but through which the blood itself must never break, or it means bleeding.

In spiritual love, the contact is purely nervous. The nerves in the lovers are set vibrating in unison like two instruments. The pitch can rise higher and higher. But carry this too far, and the nerves begin to break, to bleed, as it were, and a form of death sets in.

The trouble about man is that he insists on being master of his own fate, and he insists on *oneness*. For instance, having discovered the ecstasy of spiritual love, he insists that he shall have this all the time, and nothing but this, for this is life. It is what he calls "heightening" life. He wants his nerves to be set vibrating in the intense and exhilarating unison with the nerves of another being, and by this means he acquires an ecstasy of vision, he finds himself in glowing unison with all the universe.

But as a matter of fact this glowing unison is only a temporary thing, because the first law of life is that each organism is isolate in itself, it must return to its own isolation.

Yet man has tried the glow of unison, called love, and he *likes* it. It gives him his highest gratification. He wants it. He wants it all the time. He wants it and he will have it. He doesn't want to return to his own isolation. Or if he must, it is only as a prowling beast returns to its lair to rest and set out again.

This brings us to Edgar Allan Poe. The clue to him lies in the motto he chose for *Ligeia,* a quotation from the mystic Joseph Glanville:

And the will therein lieth, which dieth not. Who knoweth the mysteries of the will, with its vigour? For God is but a great Will pervading all things by nature of its intentness. Man doth not yield himself to the angels, nor unto death utterly, save only through the weakness of his feeble will.

It is a profound saying: and a deadly one.

Because if God is a great will, then the universe is but an instrument.

I don't know what God is. But He is not simply a will. That is too simple. Too anthropomorphic. Because a man wants his own will, and nothing but his will, he needn't say that God is the same will, magnified *ad infinitum*.

For me, there may be one God, but He is nameless and unknowable.

For me, there are also many gods, that come into me and leave me again. And they have very various wills, I must say.

But the point is Poe.

Poe had experienced the ecstasies of extreme spiritual love. And he wanted those ecstasies and nothing but those ecstasies. He wanted that great gratification, the sense of flowing, the sense of unison, the sense of heightening of life. He had experienced this gratification. He was told on every hand that this ecstasy of spiritual, nervous love was the greatest thing in life, was life itself. And he had tried it for himself, he knew that for him it *was* life itself. So he wanted it. And he *would have* it. He set up his will against the whole of the limitations of nature.

This is a brave man, acting on his own belief, and his own experience. But it is also an arrogant man, and a fool.

Poe was going to get the ecstasy and the heightening, cost what it might. He went on in a frenzy, as characteristic American women nowadays go on in a frenzy, after the very same thing: the heightening, the flow, the ecstasy. Poe tried alcohol, and any drug he could lay his hand on. He also tried any human being he could lay his hands on.

His grand attempt and achievement was with his wife; his cousin, a girl with a singing voice. With her he went in for the intensest flow, the heightening, the prismatic shades of ecstasy. It was the intensest nervous vibration of unison, pressed higher and higher in pitch, till the blood vessels of the girl broke, and the blood began to flow out loose. It was love. If you call it love.

Love can be terribly obscene.

It is love that causes the neuroticism of the day. It is love that is the prime cause of tuberculosis.

The nerves that vibrate most intensely in spiritual unisons are the sympathetic ganglia of the breast, of the throat, and the hind brain. Drive this vibration over-intensely, and you weaken the sympathetic tissues of the chest—the lungs—or of the throat, or of the lower brain, and the tubercles are given a ripe field.

But Poe drove the vibrations beyond any human pitch of endurance. Being his cousin, she was more easily keyed to him.

*   *   *

The other great story linking up with this group is *The Fall of the House of Usher*. Here the love is between brother and sister. When the self is broken, and the mystery of the recognition of *otherness* fails, then the longing for identification with the beloved becomes a lust. And it is this longing for identification, utter merging, which is at the base of the incest problem. In psychoanalysis almost every trouble in the psyche is traced to an incest-desire. But it won't do. Incest-desire is only one of the modes by which men strive to get their gratification of the intensest vibration of the spiritual nerves, without any resistance. In the family, the natural vibration is most nearly in unison. With a stranger, there is greater resistance. Incest is the getting of gratification and the avoiding of resistance.

The root of all evil is that we all want this spiritual gratification, this flow, this apparent heightening of life, this knowledge, this valley of many-coloured grass, even grass and light prismatically decomposed, giving ecstasy. We want all this *without resistance*. We want it continually. And this is the root of all evil in us.

We ought to pray to be resisted and resisted to the bitter end. We ought to decide to have done at last with craving.

The motto to *The Fall of the House of Usher* is a couple of lines from Béranger.

> *Son coeur est un luth suspendu;*
> *Sitôt qu'on le touche il résonne.** *

We have all the trappings of Poe's rather overdone, vulgar fantasy.

I reined my horse to the precipitous brink of a black and lurid tarn that lay in unruffled lustre by the dwelling, and gazed down—but with a shudder even more thrilling than before—upon the remodelled and inverted images of the grey sedge, and the ghastly tree-stems, and the vacant and eye-like windows.

The House of Usher, both dwelling and family, was very old. Minute fungi overspread the exterior of the house, hanging in festoons from the eves. Gothic archways, a valet of stealthy step, sombre tapestries, ebon black floors, a profusion of tattered and antique furniture, feeble gleams of encrimsoned light through latticed panes, and over all "an air of stern, deep, irredeemable gloom"—this makes up the interior.

The inmates of the house, Roderick and Madeline Usher, are the last remnants of their incomparably ancient and decayed race. Roderick has the same large, luminous eye, the same slightly arched nose of delicate Hebrew model, as characterized Ligeia. He is ill with the nervous malady of his family. It is he whose nerves are so strung that they vibrate to the unknown quiverings of the ether. He, too, has lost his self, his living soul, and become a sensitized instrument of the external influences; his nerves are verily like an aeolian harp which must vibrate. He lives in "some struggle with the grim phantasm,

---

*His heart is a suspended lute;
As soon as it is touched, it resounds. [Editor]

Fear," for he is only the physical, post-mortem reality of a living being.

It is a question how much, once the true centrality of the self is broken, the instrumental consciousness of man can register. When man becomes self-less, wafting instrumental like a harp in an open window, how much can his elemental consciousness express? The blood as it runs has its own sympathies and responses to the material world, quite apart from seeing. And the nerves we know vibrate all the while to unseen presences, unseen forces. So Roderick Usher quivers on the edge of material existence.

It is this mechanical consciousness which gives "the fervid facility of his impromptus." It is the same thing that gives Poe his extraordinary facility in versification. The absence of real central or impulsive being in himself leaves him inordinately mechanically sensitive to sounds and effects, associations of sounds, associations of rhyme, for example —mechanical, facile, having no root in any passion. It is all a secondary, meretricious process. So we get Roderick Usher's poem, *The Haunted Palace,* with its swift yet mechanical subtleties of rhyme and rhythm, its vulgarity of epithet. It is all a sort of dream-process, where the association between parts is mechanical, accidental as far as passional meaning goes.

Usher thought that all vegetable things had sentience. Surely all material things have a *form* of sentience, even the inorganic: surely they all exist in some subtle and complicated tension of vibration which makes them sensitive to external influence and causes them to have an influence on other external objects, irrespective of contact. It is of this vibration or inorganic consciousness that Poe is master: the sleep-consciousness. Thus Roderick Usher was convinced that his whole surroundings, the stones of the house, the fungi, the water in the tarn, the very reflected image of the whole, was woven into a physical oneness with the family, condensed, as it were, into one atmosphere—the special atmosphere in which alone the Ushers could live. And it was this atmosphere which had moulded the destinies of his family.

But while ever the soul remains alive, it is the moulder and not the moulded. It is the souls of living men that subtly impregnate stones, houses, mountains, continents, and give these their subtlest form. People only become subject to stones after having lost their integral souls.

In the human realm, Roderick had one connection: his sister

Madeline. She, too, was dying of a mysterious disorder, nervous, cataleptic. The brother and sister loved each other passionately and exclusively. They were twins, almost identical in looks. It was the same absorbing love between them, this process of unison in nerve-vibration, resulting in more and more extreme exaltation and a sort of consciousness, and a gradual break-down into death. The exquisitely sensitive Roger, vibrating without resistance with his sister Madeline, more and more exquisitely, and gradually devouring her, sucking her life like a vampire in his anguish of extreme love. And she asking to be sucked.

Madeline died and was carried down by her brother into the deep vaults of the house. But she was not dead. Her brother roamed about in incipient madness—a madness of unspeakable terror and guilt. After eight days they were suddenly startled by a clash of metal, then a distinct, hollow metallic, and clangorous, yet apparently muffled, reverberation. Then Roderick Usher, gibbering, began to express himself: *"We have put her living into the tomb!* Said I not that my senses were acute? I *now* tell you that I heard her first feeble movements in the hollow coffin. I heard them—many, many days ago—yet I dared not—*I dared not speak."*

It is the same old theme of "each man kills the thing he loves." He knew his love had killed her. He knew she died at last, like Ligeia, unwilling and unappeased. So, she rose again upon him.

But then without those doors there *did* stand the lofty and enshrouded figure of the Lady Madeline of Usher. There was blood upon her white robes, and the evidence of some bitter struggle upon every portion of her emaciated frame. For a moment she remained trembling and reeling to and fro upon the threshold, then, with a low moaning cry, fell heavily inward upon the person of her brother, and in her violent and now final death-agonies bore him to the floor a corpse, and a victim to the terrors he had anticipated.

It is lurid and melodramatic, but it is true. It is a ghastly psychological truth of what happens in the last stages of this beloved love, which cannot be separate, cannot be isolate, cannot listen in isolation to the isolate Holy Ghost. For it is the Holy Ghost we must live by. The next era is the era of the Holy Ghost. And the Holy Ghost speaks individually inside each individual: always, for ever a ghost. There is no manifestation to the general world. Each isolate individual listening in isolation to the Holy Ghost within him.

The Ushers, brother and sister, betrayed the Holy Ghost in themselves. They would love, love, love, without resistance. They would love, they would merge, they would be as one thing. So they dragged each other down into death. For the Holy Ghost says you must *not* be as one thing with another being. Each must abide by itself, and correspond only within certain limits.

The best tales all have the same burden. Hate is as inordinate as love, and as slowly consuming, as secret, as underground, as subtle. All this underground vault business in Poe only symbolizes that which takes place *beneath* the consciousness. On top, all is fair-spoken. Beneath, there is awful murderous extremity of burying alive.

* * *

So the mystery goes on. La Bruyère says that all our human unhappinesses *viennent de ne pouvoir être seuls.* As long as man lives he will be subject to the yearning of love or the burning of hate, which is only inverted love.

But he is subject to something more than this. If we do not live to eat, we do not live to love either.

We live to stand alone, and listen to the Holy Ghost. The Holy Ghost, who is inside us, and who is many gods. Many gods come and go, some say one thing and some say another, and we have to obey the God of the innermost hour. It is the multiplicity of gods within us make up the Holy Ghost.

But Poe knew only love, love, love, intense vibrations and heightened consciousness. Drugs, women, self-destruction, but anyhow the prismatic ecstasy of heightened consciousness and sense of love, of flow. The human soul in him was beside itself. But it was not lost. He told us plainly how it was, so that we should know.

He was an adventurer into vaults and cellars and horrible underground passages of the human soul. He sounded the horror and the warning of his own doom.

Doomed he was. He died wanting more love, and love killed him. A ghastly disease, love. Poe telling us of his disease: trying even to make his disease fair and attractive. Even succeeding.

Which is the inevitable falseness, duplicity of art, American Art in particular.

A Key to the House of Usher

by Darrel Abel

By common consent, the most characteristic of Poe's "arabesque" tales is "The Fall of the House of Usher." It is usually admired for its "atmosphere" and for its exquisitely artificial manipulation of Gothic claptrap and décor, but careful reading reveals admirable method in the author's use of things generally regarded by his readers as mere decorative properties.

Poe insisted that the "calculating" and "ideal" faculties, far from being "at war" with each other, were complementary aspects of the creative imagination—a doctrine vigorously reasserted by critics in our own day. He further maintained that "it is an obvious rule of Art that effects should be made to spring as directly as possible from their causes." Such an emphasis on logical exactitude in the calculation of artistic effect invites inquiry into the question: How far is the unity of effect (which he called "that vital requisite of all works of Art") in one of Poe's most characteristic and successful works directly analysable into its causes? It will be seen that his effects are produced by deeper causes than has been supposed by those casual critics who believe that the horror of "The Fall of the House of Usher" is merely an adventitious product of "atmosphere."

I

Too much of the horror of the tale has usually been attributed to its setting superficially considered. But the setting does have a double importance, descriptive and symbolic. It first operates descriptively, as suggestively appropriate and picturesque background for the un-

"*A Key to the House of Usher*" *by Darrel Abel. From* The University of Toronto Quarterly, *XVIII (1949), 176–85. Copyright 1949 by* The University of Toronto Quarterly. *Reprinted by permission of The University of Toronto Press and Darrel Abel.*

folding of events. It later operates symbolically: certain features of the setting assume an ominous animism and function; they become important active elements instead of mere static backdrop.

Descriptively the setting has two uses: to suggest a mood to the observer which makes him properly receptive to the horrible ideas which grow in his mind during the action; and to supply details which reinforce, but do not produce, those ideas.

The qualities of the setting are remoteness, decadence, horrible gloom. Remoteness (and loss of feature) is suggested by details of outline, dimension, and vista. Decadence is suggested by details of the death or decrepitude of normal human and vegetable existences and constructions, and by the growth of morbid and parasitic human and vegetable existences, as well as by the surging sentience of inorganism. Gloom and despair are suggested by sombre and listless details of colour and motion (at climactic points, lurid colour and violent action erupt with startling effect from this sombre listlessness). The narrator points out in the opening passage of the tale that the gloom which invested the domain of Usher was not sublime and pleasurable (which would have made it an expression of "supernal beauty" in Poe's opinion), but was sinister and vaguely terrible.

Five persons figure in the tale, but the interest centres exclusively in one—Roderick Usher. The narrator is uncharacterized, undescribed, even unnamed. (I shall call him Anthropos, for convenient reference.) In fact, he is a mere point of view for the reader to occupy, but he does lend the reader some acute, though not individualizing, faculties: five keen senses which shrewdly perceive actual physical circumstances; a sixth sense of vague and indescribable realities behind the physical and apparent; a clever faculty of rational interpretation of sensible phenomena; and finally, a sceptical and matter-of-fact propensity to mistrust intuitional apprehensions and to seek natural and rational explanations. In short, he is an habitual naturalist resisting urgent convictions of the preternatural.

The doctor and valet are not realized as characters; they are less impressive than the furniture; and Anthropos sees each only once and briefly. No duties requiring the attendance of other persons are mentioned, so our attention is never for a moment diverted from Roderick Usher. His sister Madeline's place in the story can best be explained in connection with comment on Usher himself.

The action of the story is comparatively slight; the energetic symbolism, to be discussed later, accomplishes more. Anthropos arrives at the House of Usher, and is conducted into the presence of his host. Usher has invited Anthropos, a friend of his school-days, in the hope that a renewal of their association will assist him to throw off a morbid depression of spirits which has affected his health. Anthropos is shocked at the ghastly infirmity of his friend. He learns that Madeline, Roderick's twin and the only other living Usher, is near death from a mysterious malady which baffles her physicians. Presently she dies and Roderick Usher, fearing that the doctors who had been so fascinated by the pathology of the case might steal her body from the grave, places it in a sealed coffin in a subterranean vault under the House of Usher. Anthropos assists in this labour.

Immediately there is an observable increase in the nervous apprehensiveness of Roderick Usher. He finds partial relief from his agitation in the painting of horribly vague abstract pictures and in the improvisation of wild tunes to the accompaniment of his "speaking guitar." For seven or eight days his apprehensiveness increases and steadily communicates itself to Anthropos as well, so that, at the end of that time, a night arrives when Anthropos' state of vague alarm prevents his going to sleep. Usher enters and shows him through the window that, although the night is heavily clouded, the House of Usher's environs are strangely illuminated. Anthropos endeavours, not very judiciously, to calm him by reading aloud from a romance that might have come from the library of Don Quixote. At points of suspense in this romance, marked by description of loud noises, Anthropos fancies that he hears similar sounds below him in the House of Usher. Roderick Usher's manner, during this reading, is inattentive and wildly preoccupied; at the noisy climax of the romance Usher melodramatically shrieks that the noises outside had actually been those of his sister breaking out of the coffin in which she had been sealed alive. The door bursts open; Madeline appears and, falling forward dead in her gory shroud, carries Roderick Usher likewise dead to the floor beneath her. Anthropos rushes from the House of Usher, turning in his flight to view its shattering collapse into the gloomy tarn beneath it. How these events become invested with horror can only be understood by discerning the meanings which the symbolism of the tale conveys into them.

II

Roderick Usher is himself a symbol—of isolation, and of a concentration of vitality so introverted that it utterly destroys itself. He is physically isolated. Anthropos reaches the House of Usher after a whole day's journey "through a singularly dreary tract of country" that is recognizably the same sort of domain-beyond-reality as that traversed by Childe Roland and his medieval prototypes. Arrived at the mansion, he is conducted to Usher's "studio" "through many dark and intricate passages." And there "the eye struggled in vain to reach the remoter angles of the chamber" in which his host received him.

Usher is psychologically isolated. Although he has invited his former "boon companion" to visit and support him in this moral crisis, clearly there has never been any conviviality in his nature. "His reserve had always been habitual and excessive," and he has now evidently become more singular, preoccupied, and aloof than before. "For many years, he had never ventured forth" from the gloomy House of Usher, wherein "he was enchained by certain superstitious impressions." ("Superstitious" is the sceptical judgment of Anthropos.) Thus, although his seclusion had probably once been voluntary, it is now inescapable. His sister Madeline does not relieve his isolation; paradoxically, she intensifies it, for they are twins whose "striking similitude" and "sympathies of a scarcely intelligible nature" eliminate that margin of difference which is necessary to social relationship between persons. They are not two persons, but one consciousness in two bodies, each mirroring the other, intensifying the introversion of the family character. Further, no collateral branches of the family survive; all the life of the Ushers is flickering to extinction in these feeble representatives. Therefore no wonder that Anthropos cannot connect his host's appearance "with any idea of a simple humanity."

The isolation and concentration of the vitalities of the Ushers had brought about the decay of the line. Formerly the family energies had found magnificently varied expression: "His very ancient family had been noted, time out of mind, for a peculiar sensibility of temperament; displaying itself, through long ages, in many works of exalted art, and manifested, of late, in repeated deeds of munificent yet unobtrusive charity, as well as in a passionate devotion to the intricacies, perhaps even more than to the orthodox and easily recognizable beauties, of musical science." For all the splendid flowering of this "peculiar sensibility," its devotion to intricacies was a fatal weakness; in tending inward to more hidden channels of expression, the family sensibility had become in its current representative morbidity introverted from lack of proper object and exercise, and its only flowers were flowers of evil. It was fretting Roderick Usher to death: "He suffered much from a morbid acuteness of the senses; the most insipid food was alone endurable; he could wear only garments of a certain texture; the odors of all flowers were oppressive; his eyes were tortured by even a faint light; and there were but peculiar sounds, and these from stringed instruments, which did not inspire him with horror." These specifications detail the hyper-acuity but progressive desuetude of his five senses. The sum of things which these five senses convey to a man is the sum of physical life; the relinquishment of their use is the relinquishment of life itself. The hyper-acuity of Roderick Usher's senses was caused by the introverted concentration of the family energies; the inhibition of his senses was caused by the physical and psychological isolation of Usher. It is noteworthy that the only willing use he makes of his senses is a morbid one—not to sustain and positively experience life, but to project his "distempered ideality" on canvas and in music. This morbid use of faculties which ought to sustain and express life shows that, as Life progressively loses its hold on Roderick Usher, Death as steadily asserts it empery over him. The central action and symbolism of the tale dramatize this contest between Life and Death for the possession of Roderick Usher.

III

Some of the non-human symbols of the tale are, as has been mentioned, features of the physical setting which detach themselves from the merely picturesque ensemble of background particulars and assume symbolical meaning as the tale unfolds. They have what might be called an historical function; they symbolize what has been and is. The remaining symbols are created by the "distempered ideality" of Roderick Usher as the narrative progresses. These have prophetic significance; they symbolize what is becoming and what will be. The symbols which Usher creates, however, flow from the same dark source as the evil in symbols which exist independently of Usher: that evil is merely channelled through his artistic sensibility to find bold new expression.

All the symbols express the opposition of Life-Reason to Death-Madness. Most of them are mixed manifestations of those two existences; more precisely, they show ascendant evil encroaching upon decadent good. On the Life-Reason side are ranged the heavenly, natural, organic, harmonious, featured, active qualities of things. Against them are ranged the subterranean, subnatural, inorganic, inharmonious, vague or featureless, passive qualities of things. Although most of the symbols show the encroachment of Death-Madness on Life-Reason, two symbols show absolute evil triumphant, with no commixture of good even in decay. One of these is the tarn, a physically permanent feature of the setting; the other is Roderick Usher's ghastly abstract painting, an impromptu expression of the evil which has mastered his sensibility. There are no symbols of absolute good.

The House of Usher is the most conspicuous symbol in the tale. It displays all the qualities (listed above) of Life-Reason, corrupted and threatened by Death-Madness. It stands under the clouded heavens, but it is significantly related to the subterranean by the zigzag crack which extends from its roof (the most heavenward part of the house) to the tarn. The trees about it connect it with nature, but they are all dead, blasted by the preternatural evil of the place; the only living vegetation consists of "rank sedges" (no doubt nourished by the tarn), and fungi growing from the roof, the most heavenward part. The house is also a symbol of the organic and harmonious because it expresses human thought and design, but the structure is crazy, threatened not only by the ominous, zigzag, scarcely discernible fissure, but also by the perilous decrepitude of its constituent materials, which maintained their coherency in a way that looked almost miraculous to Anthropos: "No portion of the masonry had fallen; and there appeared to be a wild inconsistency between its still perfect adaptation of parts, and the crumbling condition of the individual stones." In the interior of the house, the furnishings seemed no longer to express the ordered living of human creatures: "The general furniture was profuse, comfortless, antique, and tattered. Many books and musical instruments lay scattered about, but failed to give any vitality to the scene." That is, the human life it expressed was not ordered and full, but scattered and tattered. The "eyelike windows," the most conspicuous feature of the house, looked vacant from without, and from within were seen to be "altogether inaccessible"; they admitted only "feeble gleams of encrimsoned light." Life and motion within the house were nearly extinct. "An air of stern, deep, and irredeemable gloom hung over and pervaded all."

Roderick Usher resembles his house. It is unnecessary to point out the ways in which a human being is normally an expression of Life-Reason—of heavenly, natural, organic, harmonious, featured, and active qualities. The Death-Madness opposites to these qualities are manifested in interesting correspondences between the physical appearance of Usher and that of his house. The zigzag crack in the house, and the "inconsistency" between its decayed materials and intact structure, are like the difficulty maintained composure of Usher. Anthropos declares: "In the manner of my friend I was at once struck with an incoherence—an *inconsistency* [my italics]; and I soon found this to arise from a series of feeble and futile struggles to overcome an habitual trepidancy—an excessive nervous agitation." The "minute fungi . . . hanging in a fine tangled web-work from the eaves" of the house have their curious counterpart, as a symbol of morbid vitality, in the hair of Usher, "of a more than web-like softness and tenuity . . . [which, as it] had been suffered to grow all unheeded, . . . floated rather than fell about the face, [so that] I could not, even with an effort, connect its Arabesque expression with any idea of simple humanity." (We are reminded of the hair reputed to grow so luxuriantly out of the heads of inhumed corpses.) Usher's organic existence and sanity seem threatened: his "cadaverousness of complexion" is conspicuous; and he not only attributes sentience to vegetable things, but also to "the kingdom of inorganization" which he evidently feels to be assuming domination over him. His most conspicuous feature was "an eye large, liquid, and luminous beyond comparison"; after Madeline Usher's death, Anthropos observes that "the luminousness of his eye had utterly gone out." It was thus assimilated to the "vacant eye-like windows" of his house. And the active qualities of Usher were also fading. We have noticed that his malady was a combined hyperacuity and inhibition of function of the five senses which maintain life and mind. Altogether, the fabric of Usher, like that of his house, exhibited a "specious totality."

The only other important mixed symbol is Usher's song of the "Haunted Palace." It is largely a contrast of before and after. Before the palace was assailed by "evil things, in robes of sorrow," it had "reared its head" grandly under the heavens:

> Never seraph spread a pinion
> Over fabric half so fair!

It displayed several of the characteristics of Life-Reason. But after the assault of "evil things," the Death-Madness qualities are triumphant. Order is destroyed; instead of

> Spirits moving musically
> To a lute's well-tuned law,

within the palace are to be seen

> Vast forms that move fantastically
> To a discordant melody.

Instead of a "troop of Echoes" flowing and sparkling through the "fair palace door," "a hideous throng rush out for ever" through the "pale door" "like a rapid ghastly river." Reason has toppled from its throne, and this song intimated to Anthropos "a full consciousness on the part of Usher of the tottering of his lofty reason upon her throne." The perceptible fading of bright features in the palace is like the fading of the features and vitality of both Usher and his house.

The principal symbols of decrepit Life-Reason having been explicated, it remains to comment on the two symbols of ascendant Death-Madness—the tarn, and Roderick Usher's madly abstract painting. These show the same qualities that we have seen evilly encroaching upon the Life-Reason symbols, but these qualities are here unmitigated by any hint or reminiscence of Life-Reason. The juxtaposition of the tarn-house symbols is crucial; the zigzag fissure in the house is an index to the source of the evil which eventually overwhelms the Ushers. The tarn is an outlet of a subterranean realm; on the surface of the earth this realm disputes dominion with the powers of heaven and wins. This subnatural realm manifests itself in the miasma that rises from the tarn.

> About the mansion and the whole domain there hung an atmosphere peculiar to themselves and their immediate vicinity—an atmosphere which had no affinity with the air of heaven, but which had reeked up from the decayed trees, and the gray wall, and the silent tarn—a pestilent and mystic vapor, dull, sluggish, faintly discernible, and leaden-hued.

This upward-reeking effluvium has its counterpart in the "distempered ideality" of Usher while he is producing his mad compositions after the death of Madeline: they are products of "a mind from which dark-

LIBRARY
ZION-BENTON TOWNSHIP HIGH SCHOOL
ZION, ILLINOIS

3/7/3

ness, as if an inherent positive quality, poured forth upon all objects of the moral and physical universe in one unceasing radiation of gloom."

"Radiation of gloom" is as interesting an idea as "darkness visible." It reminds us that another mark of this emanation of evil was lurid illumination. The feeble gleams of light that entered Usher's studio were encrimsoned. The "luminous windows" of the "radiant palace" became the "red-litten windows" of the "haunted palace." Oddly, even Usher's mad music is described in a visual figure as having a "sulphureous lustre." On the catastrophic last night of the House of Usher, the environs are at first illuminated, not by any celestial luminaries, but by the "unnatural light of a faintly luminous and distinctly visible gaseous [so our matter-of-fact Anthropos] exhalation which hung about and enshrouded the mansion." And finally, the collapse of the house is melodramatically spotlighted by "the full, setting, and blood-red moon, which now shone vividly through that once barely perceptible fissure."

Roderick Usher's dread of the "kingdom of inorganization" as a really sentient order of existence reminds us of the animate inanimation of the tarn. Activity and harmony are really related qualities; harmony is an agreeable coincidence of motions. The tarn's absolute stillness is the negation of these qualities. Water is a universal and immemorial symbol of life; this dead water is thus a symbol of Death-in-Life. It lies "unruffled" from the first, and when at last the House of Usher topples thunderously into it, to the noisy accompaniment of Nature in tumult, its waters close "sullenly and silently over the fragments." This horrid inactivity is the condition toward which Usher is tending when he finds the exercise of his senses intolerable.

The tarn is as featureless as any visible thing can be; its blackness, "unruffled lustre," and silence are like the painted "vaguenesses" at which Anthropos shuddered "the more thrillingly" because he shuddered "not knowing why." Here are blank horrors, with only enough suggestion of feature to set the imagination fearfully to work.

This leads us to the only remaining symbol of importance, Usher's terrible painting. It is more horrible than the "Haunted Palace" because, whereas the song described the lost but regretted state of lovely Life and Reason, the painting depicts Death-Madness horribly regnant, with no reminiscence of Life and Reason. The scene pictured is subterranean (Madeline's coffin was deposited in a suggestively similar vault): "Certain accessory points of the design served well to convey the idea that this excavation lay at an exceeding depth below the surface of the earth." It is preternaturally lurid: "No torch or other artificial source of light was discernible; yet a flood of intense rays rolled throughout, and bathed the whole in a ghastly and inappropriate splendor." The picture shows a lifeless scene without features —"smooth, white, and without interruption or device."

Before these remarks on the symbolism of the tale are concluded, some notice should be taken of the part which musical symbols play in it. Poe uses his favorite heart-lute image, from Béranger, as a motto:

> Son cœur est un luth suspendu;
> Sitôt qu'on le touche il résonne.*

The "lute's well-tuned law" symbolizes ideal order in the "radiant palace," and the whole of that song is an explicit musical metaphor for derangement of intellect. For Poe, music was the highest as well as the most rational expression of the intelligence, and string music was quintessential music (wherefore Usher's jangled intellect can endure only string music). Time out of mind, music has symbolized celestial order. His conception was not far from that expressed in Dryden's "Song for St. Cecilia's Day," with "The diapason closing full in Man." The derangement of human reason, then, "sweet bells jangled out of tune and harsh," cannot be better expressed than in a musical figure.

IV

I have thus tediously but by no means exhaustively exposed the filaments of symbol in "The Fall of the House of Usher" to show how much of its effect depends on the artfully inconspicuous iteration and reiteration of identical suggestions which could not operate so unobtrusively in any other way. Human actions in the story are of much less importance, but one or two events deserve notice. The depositing of Madeline's coffin in the underground vault provides Anthropos with an opportunity to compare the appearance of Roderick and Madeline Usher. She had on her face and bosom "the mockery of a faint blush" and on her lips "that suspiciously lingering smile . . . which is so terrible in death." In contrast, the "cadaverousness of complexion"

---

* See footnote on page 39. [Editor]

of Roderick Usher had been repeatedly remarked. Thus is indicated how nearly triumphant Death is in the Ushers from the moment when Anthropos first enters the house, how scarcely perceptible is the difference between a live Usher and a dead one. Consequently, Madeline's rising up from her coffin to claim her brother for death really suggests that he had mistakenly and perversely lingered among the living, that the similitude of life in an Usher was merely morbid animation. He needed only to cross a shadowy line to yield himself up to Madness and Death.

The night of catastrophe, then, witnessed this transition. The reading of "The Mad Trist" shows a mechanical, not a symbolical, correspondence between Usher's ruin and external things; it is the only piece of superimposed and unfunctional trumpery in the tale, though it does serve, perhaps, to explain and justify the suspenseful doubt and surprise of Anthropos when he hears the weird sounds of Madeline's ghastly up-rising. The storm which rages outside is not a supernatural storm, but a tumult of natural elements impotently opposing the silent and sullen powers which in that hour assert dominion over the House of Usher and draw it into their Plutonian depths.

The tragedy of Roderick Usher was not merely his fatal introversion, but his too-late realization of his own doom, the ineffectuality of his effort to re-establish connection with life by summoning to him the person most his friend. When at last he shrieks "Madman!" at this presumably sane friend, he crosses the borderline between sanity and madness. In a moment he dies in melodramatic circumstances, and immediately thereafter is carried into the tarn by the culminatingly symbolical collapse of his house.

V

It is expedient to review the impressions of Anthropos the determined doubter, who leaves the domain of Usher with a sense of supernatural fatality accomplished. Throughout the tale he scrupulously tries to find rational explanations for the horrors which agitate him. He explains his depression of spirits when he first views the House of Usher by reference to the gloomy combination of "very simple natural objects." That the tarn deepened this depression he accounted for psychologically: "The consciousness of the rapid increase of my superstition—for why should I not so term it?—served mainly to accelerate the increase itself." In the house he is puzzled to account for the fact that, although the furniture is all of a sort to which he has been accustomed throughout his life, it has an "unfamiliar" effect of gloom; and it is difficult for him to connect any "idea of simple humanity" with Usher's ghastly appearance, although he dutifully tries. He tells us that Usher "admitted" that his "superstitious impressions in regard to the dwelling which he tenanted" might be traced to "a more natural and far more palpable origin" than the malign sentience which he attributed to the place, that is to his grief at his sister's hopeless illness. The music which Usher composes during his bereavement is characterized by his common-sensible friend as distempered and perverted, and Usher himself is called a hypochondriac. The limited tolerance of Usher for sound is described in Anthropos' medical jargon as "a morbid condition of the auditory nerve." Usher's conviction of the sentience of the "kingdom of inorganization" is regarded by his friend as a pertinacious but not altogether novel delusion. Usher's agitation is partly ascribed to the influence of the fantastic literature which he reads. The sounds which interrupt the reading of "The Mad Trist" are, Anthropos thinks (before the apparition of Madeline changes his opinion), hallucinations prompted by the wild story and his own state of excited suggestibility. The lurid, upward-streaming illumination of the environs of the House of Usher on the night of catastrophe is explained as a natural phenomenon—a "gaseous exhalation." And, if we wish, we can attribute the stupendously shattering collapse of the ancient House of Usher itself to merely physical and natural causes— the violent thrust of the storm against its frail fabric and almost dilapidated structure. But, significantly, our matter-of-fact Anthropos does not suggest any natural explanation; he merely flees "aghast."

VI

"The Fall of the House of Usher" has that "vital requisite of all works of Art, Unity." It achieves unity by a concentration and remarkably subtle co-operation of carefully calculated, complex causes. Instead of the more familiar methods of realistic narrative specification and progressive logical explication, it operates through parallel symbolic suggestions. These build up to such a climax of intensity that the final shattering crash of all these piled-up effects affords a powerful release of psychological tension.

Throughout the tale, alternative explanations, natural and supernatural, of the phenomena are set forth; and we are induced, by the consistently maintained device of a common-sense witness gradually convinced in spite of his determined scepticism, to accept imaginatively the supernatural explanation.

The tale is a consummate psychological allegory which produces its intended totality of effect by perfectly unobtrusive means; there are no seams, bastings, or other interfering reminders of its being artificially put together. It is, indeed, too successful: readers take it to be all shell, and, although it irresistibly makes its intended impression, its method is so concealed that the too casual reader may take the impression to be meretricious. Those who admire Kafka only this side idolatry for employing a method which they look upon as a brilliant innovation, will find his technique anticipated in Poe without the baffling obscurities and seeming irrelevancies of Kafka: Poe's is the art which conceals art.

A Reinterpretation of "The Fall of the House of Usher"

by Leo Spitzer

"When the mill of the poet starts grinding, do not attempt to stay it; for he who understands us will also know how to forgive us."

—GOETHE

Edgar Allan Poe fares badly at the hands of contemporary critics, if we may judge from the treatment given "The Fall of the House of Usher" by such subtle commentators as Cleanth Brooks and Robert Penn Warren in their *Understanding Fiction* (New York, 1943). It may be worth while to re-analyze this little masterpiece and then to elucidate Poe's artistry still further by considering it from a comparatist viewpoint.

Let us first consider the strictures made by Messrs. Brooks and Warren against this "story of horror." "The Fall of the House of Usher," they hold, is "within limits, rather successful in inducing in the reader the sense of nightmare," but "horror for its own sake" cannot be aesthetically justified unless the horror, of true tragic impact (Macbeth, Lear), "engages our own interest." Poe's protagonist, Roderick Usher, fails to engage our imaginative sympathy; "the story lacks tragic quality, even pathos." Poe has narrowed the fate of his principal character to a "clinical case" which we the readers (and also the narrator) view from without. "Free will and rational decision" exist neither in the protagonist nor in the story. Roderick—it is his

"A Reinterpretation of 'The Fall of the House of Usher'" by Leo Spitzer. *From Leo Spitzer,* Essays on English and American Literature, *edited by Anna Hatcher (Princeton: Princeton University Press, 1962), pp. 51–66. Originally published in* Comparative Literature, *IV (1952), 351–63. Copyright 1962 by Princeton University Press. Reprinted by permission of Princeton University Press.*

story alone, not that of his sister, the lady Madeline—does not struggle as he should against the doom embodied in his decaying house. Poe has "played up" the sense of gloom excessively, no doubt because of "his own morbid interest in the story." [1]

These are severe words from sensitive writers who in the same volume justly extol Faulkner and Kafka, often for proceeding in the same manner as Poe. We are asked to admire the logical and methodical behavior of the protagonist in Faulkner's story, "A Rose for Emily," as being worth both our interest and our pity, whereas Poe's portrayal of logic and method in madness should not win our interest for Roderick. Again, while Emily's crime of murder is explained by our critics as a consequence of her isolation from the world and her disregard for the limits between reality and imagination, Roderick's action is not to be explained in their eyes by similar motives—in fact, this possibility is specifically excluded.[2] And here it is striking to note the fact that, contrary to their usual practice, Brooks and Warren fail to analyze the developments in our story in their (pseudo-) logical concatenation, but are satisfied to offer us a general statement about Roderick's "vague terrors and superstitions," in which he indulges without "real choice."

We must then follow out in detail the carefully wrought concatenation of events which Poe has achieved in our story. I would contend that, far from being only the story of Roderick Usher, our story is, as the title indicates, that of "the House of Usher" (a "quaint equivocal appellation," as Poe tells us, because it embraces both the family and the mansion of the Ushers). Roderick and his sister Madeline, both of them unmarried and childless, are the last scions of the family. Although Roderick is portrayed as the principal actor in the story and Madeline as a shadow, glimpsed passing with "retreating steps" only

---

[1] For critics who have always proclaimed the self-sufficiency of the literary work of art, this is a strange relapse into the "biographical fallacy"—premature recourse to the empirical biography of the writer *before* the literary work has been carefully analyzed.

[2] Our critics obviously appreciate, as would any American reader, the resistance to doom shown by Emily Grierson more than Roderick Usher's submissiveness to doom. But does their historical sense not tell them that Poe's attempt at exploration of the "attitude of doomedness" (or "le besoin de la fatalité," as Charles Dubos called it in Byron) was at the time a new step forward in the psychological study of hitherto neglected recesses and arcana of the human mind—an adventure, as D. H. Lawrence has said, "into vaults and cellars and horrible underground passages of the human soul" (quoted by Professor N. B. Fagin, *The Histrionic Mr. Poe*, Baltimore, 1949, p. 157)?

once before her death, Madeline is still a deuteragonist in her own peculiar right, on the same level with her brother. The fact that she is on stage only for a short time and has no lines to speak (only "a low moaning cry" at the moment of death is granted her) should not lead us to underrate her importance, given her impact in the story and the interest which is aroused precisely by her mysterious appearances.

Roderick and Madeline, twins chained to each other by incestuous love, suffering separately but dying together, represent the male and the female principle in that decaying family whose members, by the law of sterility and destruction which rules them, must exterminate each other; Roderick has buried his sister alive, but the revived Madeline will bury Roderick under her falling body. The "fall" of the House of Usher involves not only the physical fall of the mansion, but the physical and moral fall of the two protagonists. The incestuous and sterile love of the last of the Ushers makes them turn toward each other instead of mating, as is normal for man and woman, with blood not their own. Within the mansion they never leave, they live in an absolute vacuum.[3] In contrast to the gay comings and goings depicted in the poem recited by Roderick ("The Haunted Palace"), which reflects the former atmosphere of the mansion, we are shown only an insignificant valet of "stealthy step" and a suspect, cunning, and perplexed family physician with a "sinister countenance" (it is quite logical that Roderick, after the supposed death of his sister, should wish to keep her body as long as possible in the mansion, away from the family burial ground "exposed" to the outside world, away also from the indiscreet questions of the inquisitive physician).

As to Madeline, although her physical weakness is great and she is subject to catalepsy, she does resist the curse that is weighing down the family. "Hitherto she had steadily borne up against the pressure of her malady"; and at the moment of death she shows superhuman strength: "the rending of her coffin, and the grating of the hinges of

---

[3] Indeed it may be said that the invitation extended to the narrator by Roderick (the "vivacious warmth" and the "perfect sincerity" of whose greeting are stressed by the author) represents the last faint surge of vitality in Roderick—the desire (hysterical as are all his impulses; he writes the invitation in a "wildly importunate way") to fill his life with some content in anticipation of the death of his sister. That Roderick had wrestled before with the idea of death is shown by his reading habits. Among the books of his choice are those centered somehow around the concept of liberation (the *Ververt* and *Chartreuse* of Gresset, the *Journey into the Blue Distance* of Tieck, *The City of the Sun* of Campanella), including liberation in respect to sex (the *Belfagor* of Machiavelli, Pomponius Mela's writings on the "Satyrs and Aegipans").

her prison, and her struggles within the coppered archway of the vault" are compared by Roderick to "the breaking of the hermit's door, and the death-cry of the dragon, and the clangor of the shield" —the feats, that is, of the doughty knight Ethelred in the romance of chivalry being read to Roderick by the narrator. Surely "the lofty and enshrouded figure of the lady Madeline of Usher," as she is presented to us in an apotheosis of majesty in death, this female Ethelred returning, bloodstained, as a "conqueror" from *her* battle with the dragon (a battle that broke the enchantment of death), is the true male and last hero of the House of Usher, while her brother has in the end become a figure of passivity whose body is reduced to a trembling mass. If Roderick is the representative of death-in-life and of the death wish, Madeline becomes in the end the embodiment of life-in-death, of the will to live, indeed of a last, powerful convulsion of that will in the dying race of the Ushers.[4]

But what force moves Roderick to start the process of self-destruction? Terms such as "clinical case," "vague terrors and superstitions," may perhaps debar us from deeper psychological insight. From the beginning Poe has made it clear that he will deal in our story with the psychological consequences of fear. When the visitor who is telling us the story receives his first glimpse of the decaying mansion, he turns, in order to divert his attention from the sinister sight, toward the tarn, only to see, with growing anguish, the dreary building reflected in its waters (a foreshadowing of the end of the story when these waters will close over the debris of the mansion)—and he writes the following significant words: "There can be no doubt that the consciousness of the rapid increase of my superstition . . . served mainly to *accelerate* the increase itself. Such, I have long known, *is the paradoxical law* of all sentiments having *terror* as a basis." [5] The psycho-

---

[4] Madeline, in spite of her significant part in the plot, is seen by the narrator and presented to the reader only as a picture, the picture of a young woman dying at the acme of her beauty—a motif hallowed by the tradition of Renaissance literature (Poliziano, Lorenzo il Magnifico, Ronsard, Garcilaso).

[5] Italics mine.—Our critics do not take up the question of the role of this visitor who is the narrator (except to deplore his lack of sympathy, understanding, and information in regard to Roderick, whom he seems to treat, according to our critics, as "a clinical case"). But his function is not that of interpreting Roderick to us, of making us "take him seriously as a real human being," but of serving to objectify the *fears* harbored by Roderick. When a person mainly rational and scientific in his approach to occult phenomena (note his remark about the "merely electrical phenomena" in which Roderick sees "appearances"), who is able to recount the events he has witnessed with such elaborate detail and such poise, is "contagiously infected" by the atmosphere of the mansion and by Roderick's

logical law formulated here by Poe (fear increased by consciousness of fear) is valid especially for the monomaniac Roderick, who, throughout the story, is conscious of his "folly." At the beginning he explains to the narrator: "I shall perish . . . I *must* perish in the deplorable folly . . . I have, indeed, no abhorrence of danger, except in its absolute effect—in terror . . . I feel that the period will sooner or later arrive when I must abandon life and reason together, in some struggle with the grim phantom, FEAR." And at the end he is described as "a victim to the terrors he had *anticipated.*" Fear is indeed a passion or a hysteria that accelerates and anticipates. What our story "teaches" —and I wonder why our critics disregarded the capitalized word FEAR which should have suggested to them a "lesson"—is that fear, by anticipating terrible events, has a way of bringing about prematurely those very events. And, since we all are subject to fears, I do not understand how Poe's story should be lacking in general human interest.

Roderick fears the death of Madeline because this "would leave him (him, the hopeless and the frail) the last of the ancient race of the Ushers." The degeneracy of that race manifests itself in the overconsciousness of approaching extinction. And it is this fear that makes him see, in the figure immobilized by catalepsy, his sister dead—whom he then buries with hysterical haste. "Is she not hurrying to upbraid me for my haste?" he says when he hears her come back from the vault; fear has made him both anticipate and precipitate her death. There is no "vagueness" in Roderick's fears; they are clear-cut and precisely outlined in the text.

The narrator is no less careful to present from the beginning the peculiar state of Roderick's nervous agitation, which varies from indecision to energetic vivacity and concentration—a change, so the author tells us, like that from lethargy to most intense excitement in the opium eater. This manic-depressive state of his will culminates in his precipitate action. Similarly, Roderick's behavior before and after that action is motivated by what we would call today his schizoid nature; with him nerves and intellect act separately, though not unconnectedly. The crucial action is brought about by his intellect; but after he has buried Madeline alive he will be only a victim of his nerves. It is, however, his nerves that have from the beginning influenced his intellect. Suffering as he does from a "morbid acuteness of

"vague terrors," these terrors become thereby less vague and acquire objective reality. It was in order to make real, not Roderick himself, but Roderick's fears that Poe introduced the narrator into his story.

the senses" (why do our critics omit the important motto from Béranger: "Son cœur est un luth suspendu; Sitôt qu'on le touche il résonne"?),* from a nervousness that does not tolerate accumulation of sensuous detail, he is necessarily driven (especially in his artistic productions) toward "pure abstractions," "distempered ideality," nakedness of . . . design." And, since death represents the ultimate of abstraction, the zero degree of concrete reality, we are not astonished by the character of the picture he shows the narrator, a picture designed by an abstractionist *avant la lettre,* in which anticipation of death and nakedness of design converge: it shows the interior of an immensely long subterranean vault, without outlet, bathed in ghastly light. This is obviously the intellectual pattern that will materialize later in the burial of his sister in that donjon-keep which is "without means of admission for light" (until the moment when it will be invaded by the torches of the two men) and is situated "at great depth" and reached through a long, coppered archway.

The abstract pattern as it offered itself to the erratic mind of the amateur artist is acted out in reality by Roderick in a sudden move of concentrated energy.[6] But after the terrible deed has been accomplished, the morbidly acute nerves and senses take exclusive possession of the schizoid, polarized of course around the idea of death. When

---

* See footnote on page 39. [Editor]

[6] In the whole description of the mansion there is a pattern of "black-white" color arrangement (black, oaken, or ebony floors, dark draperies vs. the white enshrouded figure of lady Madeline which detaches itself from the somber background), and this contrast itself is in contrast with the gay, warm colors that once had made splendid our mansion: "in the greenest of our valleys"—"banners yellow, glorious, golden"·—"all with pearl and ruby glowing." Black and white, the shades used more in drawing than in painting, are obviously related to the "abstractionism" of the protagonist—and perhaps to Poe's imagination itself, which was excellently characterized by a French critic in 1856 (quoted by Lemonnier, *Edgar Poe et la critique française,* 1928, p. 285) in terms of the *décor* we find in our story; Poe's imagination hovers "dans des régions vagues, où luttent sans cesse le rayon et l'obscurité. Un pas de plus vers la lumière et vous aurez le génie; vers les ténèbres et vous aurez la folie. Entre deux, c'est . . . un je ne sais quoi semblable à ces lampes que l'on porte avec soi dans les souterrains et les mines, et dont la lueur tremblotante dessine sur les parois de capricieuses arabesques . . ." ["in undefined places where sunlight and darkness were in constant struggle. One step closer to the light and you have genius; closer to the dark and you have madness. Between the two there is an indefinable something that is most like those lamps men take with them underground into mines, and whose trembling glimmer traces capricious arabesques on the rock walls . . ." Editor] Since Poe delights in describing the penumbra of the human mind in which light and darkness are capriciously mixed, we may perhaps assume that the pattern "black-white" prevailing in his *décors* is given by a more basic intellectual pattern.

the storm rises, the portent of the fall of the mansion, it is his visual sense[7] that is stimulated: "there was a mad hilarity in his *eyes*"; he looks out toward the cloud formation which presages death. "And you have not *seen* it?" he asks the narrator, who answers: "You must not—you shall not *behold* this!"—"this" being "a faintly luminous and distinctly visible gaseous exhalation which hung about and enshrouded the mansion"—obviously an adumbration of the "enshrouded" figure of Madeline which we are to see later, and the signal to us that the House of Usher is giving up its soul. Later it is the auditory sense that predominates in Roderick; he is now able to detect every slightest sound accompanying Madeline's revival ("Said I not that my senses were acute?")—an event which occurs at the very moment the narrator is reading the Ethelred romance, so that the reading is accompanied by the sounds of Madeline's escape from the tomb which seem strangely to harmonize with the events of the romance. While Roderick hears only the sounds in the Ethelred romance that correspond to those caused by Madeline, these sounds themselves have another meaning in the story of Madeline. They spell her victory over the dragon of death, whereas Roderick is the embodiment of pure passive sentience; it is as if by the intensity of his feeling he had succeeded in conjuring up her presence and thus broken the spell of death, though in reality it is Madeline who has wrought her own liberation (it is she who has slain the dragon, whose fangs recede from its prey). A gust of wind opens the doors to disclose her majestic figure (note in the description of this scene the "as if"):

> As if in the superhuman energy of his utterance there had been found the potency of a spell, the huge antique panels . . . drew slowly back, upon the instant, their ponderous and ebony jaws. It was the work of the rushing gust—but then without those doors there *did* stand the lofty and enshrouded figure of the lady Madeline of Usher.

We have already noted that at the end the house gives up its soul before its actual "fall"—as though it were a human being. Our commentators point out the continuous correspondences in the descriptions of Roderick and of the mansion:

> The house itself gets a peculiar *atmosphere* . . . from its ability apparently to defy reality: to remain intact and yet to seem completely

---

[7] We notice that the "luminousness" of Roderick's eyes disappears after his horrible deed has been achieved: at the time, that is, when he is reduced to pure sentience (without power of ratiocination).

decayed in every detail. By the same token, Roderick has a wild vitality . . . which itself springs from the fact that he is sick unto death. Indeed, Roderick Usher is more than once in the story compared to the house, and by more subtle hints, by implications of descriptive detail, throughout the story, *the house is identified with its heir and owner* [italics mine]. For example, the house is twice described as having "vacant eyelike windows"—the house, it is suggested, is like a man. Or, again, the mad song, which Roderick Usher sings with evident reference to himself, describes a man under the *allegory* . . . of a house.

Such parallelisms[8] belong to the inner texture of the story. On the one hand, Roderick himself explains to the narrator the fact that

> He was enchanted by certain superstitious impressions in regard to the dwelling which he tenanted, and whence, for many years, he had never ventured forth . . . an influence which some peculiarities in the mere form and substance of his family mansion had, by dint of long sufferance . . . obtained over his spirit—an effect which the physique of the gray walls and turrets, and of the dim tarn into which they all looked down, had, at length, brought upon the *morale* of his existence.

On the other hand, he expresses his belief in the "sentience," not only of all vegetable things, but also of "the gray stones of the home of his forefathers" (as well as of the fungi which covered them and of the decayed trees which stood around), and he sees evidence of this sentience in the influence they have had on the destinies of his family and himself. In Roderick Usher's world the differences between the human (animal), vegetable, and the mineral kingdoms are abolished. Plants and stones are sentient, human beings have a plant or animal quality (the influence of plant life on him seems to be reflected by his silken hair—"as, in its wild gossamer texture, it floated rather than fell about the face, I could not . . . connect its arabesque expression with any idea of simple humanity"), and Madeline's youthful body is buried by her brother among the stones of the vault. Life and tomb, death and fall, are one in that strange world. Obviously one cannot ask of Roderick any struggle against his *ambiente* or any choice of

[8] One could add here the similar parallelism between, on the one hand, the description of the house as reminiscent of "the after-dream of the reveller upon opium" and of the vapor reeking up from the tarn as "leaden-hued" and, on the other, the characterization of Roderick's way of speaking as "that . . . hollow-sounding enunciation—that *leaden*, self-balanced, and perfectly modulated guttural utterance, which may be observed in . . . the irreclaimable *eater of opium*, during the periods of his most intense excitement."

another, since he is part and parcel of this *ambiente*; since he *is* himself plant and stone (and, thinking only of stone and tomb, must bury his sister alive). It is precisely the doom of Roderick—this man with the receding chin "speaking . . . of a want of moral energy"—that he has been "eaten up by his *ambiente*" (an expression of Dostoevsky's: *sreda zaela ego*). He may crave momentarily for liberation (as is suggested by his symbolic gesture of opening the window to the storm and his "mad hilarity" at the approach of death); but none the less he knows that his life is sealed within the mansion and the attraction of the subterranean vault proves irresistible. Indeed, his proclivity for the subterranean seems to be shared by the house itself, which in the end will be buried underground (and which at the beginning had appeared reflected in the tarn as if doomed to fall therein).

The result of the interpenetration between the *ambiente* and the inhabitants of the house ("the perfect keeping of the character of the premises with the accredited character of the people") is what Poe calls "atmosphere" and describes in atmospheric terms. For us the term "atmosphere" in its metaphoric meaning is trivial, but from Poe's words we gather that he wishes the term to be understood not only metaphorically but in its proper physical sense[9] as well:

> I had so worked upon my imagination [says the narrator] as *really* to
> believe that about the whole mansion and domain there *hung an
> atmosphere* peculiar to themselves and their immediate vicinity: an
> atmosphere which had no affinity with the air of heaven, but which had
> reeked up from the decayed trees, and the gray wall, and the silent

[9] The dictionaries inform us that this term, coined in the Neo-Latin of 17th century physicists and applied to "the ring or orb of vapour or 'vaporous air' supposed to be exhaled from the body of a planet, and so to be a part of it, which the *air* itself was not considered to be," then extended to the portion of air supposed to be in the planet's sphere of influence, then to "the aeriform environment of the earth," was also used in the 18th century for the sphere within which the attractive force of the magnet or the electrifying force acts (what Faraday later called the "field"). The metaphorical sense, "surrounding mental or moral element, environment," is first attested in English in 1797–1803 ("an extensive atmosphere of consciousness"; cf. Scott, 1828: "He lives in a perfect atmosphere of strife, blood and quarrels"), and in German even earlier (1767 with Herder, "die Atmosphäre der Katheder"; cf. Schiller, 1797: [rhythm is] "die Atmosphäre für die poetische Schöpfung"). The German term *Dunstkreis*, which is the translation of "atmosphere," is used by Goethe in *Faust*, 1 19, lines 2669–71, when Mephistopheles leaves Faust in Gretchen's room "to satiate himself with its [sensuous] atmosphere" ("Indessen könnt ihr ganz allein / An aller Hoffnung künft'ger Freuden / In ihrem Dunstkreis satt euch weiden").

tarn; a *pestilent and mystic vapor,* dull, sluggish, *faintly discernible,* and leaden-hued. [italics mine]

In Roderick's room the narrator feels that he *"breathed an atmosphere* of sorrow. An *air* of stern, deep, and irredeemable gloom *hung* over and pervaded all." Roderick himself speaks of "the gradual yet certain *condensation* of an atmosphere of their own about the waters and the walls" which have made him what he is. Conversely *"darkness,* as if an inherent positive quality, *poured forth* [from Roderick's mind] upon all objects of the moral and physical universe, in one unceasing *radiation of gloom."* The atmosphere, emanating from the totality of the objects and of the human being surrounded by them, is perceptible in terms of light and darkness; and this atmosphere of "radiation of gloom" is what I have called the "soul" of the mansion, which gives itself up at the end in the form of "a faintly luminous and distinctly visible gaseous exhalation which hung about and enshrouded the mansion." We should also remember "the blood-red moon" shining through the zigzag rift of the house, the light over Roderick's picture of the subterranean vault, the luminous quality of his eyes in the midst of the death pallor of his face, etc. Thus "atmosphere" is with Poe a sensuously (optically[10]) perceptible manifestation of the sum total of the physical, mental, and moral features of a particular environment and of the interaction of these features.

It is my conviction that we cannot understand the achievement of Poe unless we place his concept of "atmosphere" within the framework of ideas concerning *milieu* and *ambiance* which were being formulated at his time. As I have shown in *Essays in Historical Semantics* (New York, 1948), the terms *(circumambient) air, ambient medium, milieu (ambiant), ambiance, ambiente, environment,* etc., are all reflections of an ultimately Greek concept, τό περιέχον, which represented either the air, or space, or the World Spirit in which a particular object or being was contained; and that precisely in the third decade of the 19th century, in consequence of the biological research of Geoffroy Saint-Hilaire on the action of the environment,

[10] It may be remarked that, in the attestations given by Schulz-Basler for German *Atmosphäre,* it is rather the olfactory element that is stressed. The citations offered by the *NED* show already such a conventional use of our term that it is difficult to tell what was the original sensuous emphasis; the *NED* is not as explicit in this respect as one would wish it to be. Unless evidence to the contrary is discovered, I would surmise that Poe's concentration on the *visible* aspect of "atmosphere" is his peculiar contribution.

the term *milieu* (*ambiant*) was applied to sociology by Comte and to
fiction by Balzac, who liked to be considered a sociologist. The theory
of the time was that the organic being must be explained by the
environment just as the environment bears the imprint of this being.
Balzac, who in 1842, in the preface to his *Comédie humaine,* used the
term *milieu* in the sense in which Taine later developed it, wrote in
Le Père Goriot (begun in 1834) his description of the owner of the
Pension Vauquer:

> . . . La face vieillotte . . . ses petites mains potelées, sa personne dodue
> comme un rat d'église . . . *sont en harmonie* avec cette salle où suinte
> le malheur, où s'est blottie la spéculation, et dont Mme Vauquer respire
> l'air chaudement fétide sans en être écœurée . . . toute sa personne *ex-*
> *plique* la pension, comme la pension *implique* sa personne . . . l'em-
> bonpoint blafard de cette petite femme est *le produit de cette vie,* comme
> le typhus est la conséquence des exhalaisons d'un hôpital. Son jupon de
> laine tricotée . . . *résume* le salon, la salle à manger, le jardinet, *annonce*
> la cuisine et *fait pressentir les pensionnaires.** [italics mine]

The last sentence prepares us for the boarder who is to be the protagonist of the story, old Goriot, who is to be thought of as potentially
present, with all his lack of dignity and his frustration, in the slovenly
petticoat of Madame Vauquer. With 19th century determinism, mankind has developed far from the harmoniousness of Greek thought as
expressed in the idea of τό περιέχον; man is now embedded in a milieu
which may enclose him protectively like a shell, but may also represent
his doom and weigh him down with its unshakable reality.

Placed against this background, "The Fall of the House of Usher"
will appear to us as a poetic expression of sociological-deterministic
ideas which were in the air in 1839, the date when Poe wrote this
story. Indeed at one point Poe has Roderick summarize current environmental theory: ". . . an influence which some peculiarities in the
mere form and substance of his family mansion had, by dint of long
sufferance, he said, obtained over his spirit—an effect which the
physique of the gray walls . . . had, at length, brought about upon

---

* ". . . Her wizened face . . . her chubby little hands, her figure plump like a
churchmouse . . . *are in harmony* with that room, oozing misfortune, rife with
calculation, whose warm, fetid air Mme Vauguer breathes without becoming
nauseated . . . everything in her appearance *explains* the boardinghouse, just as
the house implies her appearance . . . this little woman's pallid stoutness is *the
result of her life here,* just as typhus comes from the exhalations of a hospital.
Her knitted woolen petticoat . . . *sums up* the living room; the dining room and
the little garden *foretell* the cooking and *foreshadow the boarders.* [Editor]

the *morale* of his existence." From this scientific theory Poe distills the poetic effect—just as he does when exploiting, for artistic purposes, contemporary theories of hypnotism, phrenology, and "the sentience of things." Our story is determinism made poetic, "atmospheric." [11] To ask Roderick to "resist" his environment when his character is meant to be the poetic embodiment of determinism is not consonant with the historical understanding of the climate of the story written in 1839—the story reflects what has been correctly called "le réalisme des romantiques."

It is, of course, not by chance that Poe insists on "atmosphere" in his story; he is describing an environment, not realistically as did Balzac, but "atmospherically." [12] We are not offered a description of the petticoat of the lady Madeline or of the thousand other details of the sort that serve to substantiate the heavy, oppressive, petty-bourgeois atmosphere of the Pension Vauquer, but only those details which are strictly connected with the main motif—the gossamer-like hair of Roderick, the blush on the cheek of the cataleptic Madeline, the vault and the archway, etc. It is as though the author, himself akin to his Roderick, had elaborated his story in terms of "abstract design" served by "acuity of senses"; [13] in fashioning the environment of his story he

[11] In the theoretical appendix to *Understanding Fiction* our critics take up the term "atmosphere" to tell us, on the one hand, that "The Fall of the House of Usher" is an "atmosphere story" (as opposed to a plot, a character, a theme story), a story that is, containing a considerable element of description, especially description intended to evoke a certain mood; on the other, that every story (not only Poe's "atmosphere story") possesses a certain atmosphere, which is the product of "the nature of the plot, of setting, of character delineation, of style and symbolism, of the very rhythms of the prose." But there is also the "atmosphere of an environment," exemplified in the description of the House of Usher (which description indeed produces a certain "mood") and defined by Poe in terms of the environmentalism of his time ("the perfect keeping of the character of the premises with the accredited character of the people").

[12] Professor Auerbach, commenting in his book *Mimesis* (p. 406) on the passage in *Le Père Goriot* discussed above, speaks of its "atmospheric realism" (that is, the realism with which the general *ambiente* of the Pension Vauquer is evoked). I have been using the term "atmospheric" somewhat differently as meaning "*only* atmospheric" description, description rendering only the atmosphere.

[13] This is exactly what endeared him to Baudelaire: see Professor Peyre's *Connaissance de Baudelaire*, Paris, 1951, p. 111: "les deux hommes avaient en partage un curieux mélange de traits émotifs et de traits intellectuels, une sensualité capable de s'élancer vers les régions supérieures à l'air raréfié, et une puissance d'analyste et de logicien abstrait rare chez les poètes." ["the two men shared a curious mixture of emotional and intellectual traits: a sensuality which could rise to higher levels and a more rarified atmosphere; and an analytical and logical power seldom found in poets." Editor]

has proceeded deductively,[14] starting from the concept of mad fear
and giving it sensuous detail only insofar as the senses are stimulated
by this madness. And we may suspect that Poe indulged in the description of the monomania of fear precisely because this offered him
patterns entirely intellectual, leading away from actual life and super-
imposing upon it another reality—that of madness. It is a remarkable
feature of his romantic realism that Poe can accept environmentalism

---

[14] This statement was corroborated by Poe himself (in 1842); see Fagin, *The
Histrionic Mr. Poe,* p. 163: "A skillful literary artist has constructed a tale. If
wise, he has not fashioned his thoughts to accommodate his incidents; but having
conceived, with deliberate care, a certain unique or single *effect* to be wrought
out, he then invents such incidents—he then combines such events as may best
aid him in establishing this preconceived effect." I do not agree with the opinion
expressed in Malcolm Cowley's otherwise excellent article (in *New Republic,* Nov. 5,
1945) that the idea of Poe just quoted should be paralleled with the American
absorption with mechanical devices, as illustrated by the engineering achievements
of his time; for we find the same emphasis on literary technique for the purpose of
attaining a specific effect in the ancients, in Goethe, in Valéry. And it would seem
paradoxical to condemn Poe (the archenemy of industrial progress), on the basis
of the ambiguity contained in the word "technique," for his "engineering talent in
poetry and fiction." A semantic fallacy is involved also in Professor Fagin's conception of the "histrionic" element in Poe. Fagin makes the passage quoted above,
with its emphasis on skillfully contrived novelistic plots, on "effect," serve the
quite gratuitous theory that Poe's genius is to be "explained" by a histrionic,
effect-seeking talent which was not allowed to come to fruition on the stage proper
in a man who wished to remain a "Mr. Poe": Poe has "somehow . . . missed his
true vocation and destiny" (p. 2). Since fiction tends in general toward drama (as
Henry James has taught us) and since the short story is the most dramatically concentrated form of fiction, it is not difficult for Mr. Fagin to represent the writing
of a story in terms of an analogy with the production of a play, and to single out
in Poe's stories the theatrical character of his plots, settings, lighting effects (even
"area-lighting"), stage props ("Max Reinhardtish at its lushest—and worst"), sound
effects (in "The Fall of the House of Usher": "Melodrama? Grand Guignol stuff?
Perhaps. But . . . how superbly staged!") synchronization, dénouements (similar to
those of the Greek tragedy), character delineation, and the rest ("his stories are
frequently dramatic productions in which Poe displays his craftsmanship as bard,
playwright, stage-designer, electrician, actor, elocutionist, and above all, director"—
p. 207). Apart from the fact that Mr. Fagin must confess that Poe as a playwright
(in his *Politian*) was a total failure, and apart from the more general experience
that any explanation of effect by defect, of genius by frustration, is a psychological
error, the main fallacy in such reasoning is that, through the use of the metaphorical *word* "histrionic," the fundamental difference between the art of writing
a dramatic story and the art of the stage is blurred. With the same reasoning the
author of the *Commedia* could be represented as a frustrated producer. To take
only the lighting effects (and God knows that Dante was a past master in those!),
Fagin overlooks the essential difference between lighting effects *suggested by words*
to the imagination of the reader and the material light actually released on a stage
before our eyes.

when it borders on irreality, whereas in Balzac the irreality of his monomaniacs grows out of earthy realism. With Balzac we see the solid ground on which the pyramid of his novels rests, with Poe only the peak of the pyramid which is bathed in the rarefied atmosphere of ratiocination. Our critics are wrong in reading Poe only "emotionally," not "conceptually."

If we compare both Balzac and Kafka with Poe, we find that the first two, while having an environmental realism in common which distinguishes their technique from Poe's atmospheric description, differ totally from each other in that Balzac's is an empirical (inductive) realism while Kafka's is a deductive realism (there is a deductive element also in Poe). Thus, environmentalism is portrayed (1) with empirical (factual) realism by Balzac, (2) with deductive realism by Kafka ("as-if realism"), and (3) with deductive irrealism by Poe ("only atmospheric" realism).

If our critics fail to appreciate "The Fall of the House of Usher" while admiring Kafka's story "In the Penal Colony" (and probably Balzac's *Le Père Goriot*), the reason may be that with Poe's atmospheric environmentalism (which is realistic only insofar as he makes the atmosphere real) the details of the description are inspired, not by realistic observation of actual contemporary mansions, but by reminiscences of conventional *literary* patterns outmoded at his time (the haunted castle of Mrs. Radcliffe, etc.). It is surely not in his choice of such hackneyed stage props that Poe's inventiveness lies, but in the arrangement to which he subjects them in order to form patterns of intellectual design.

As for Kafka, though his story, "In the Penal Colony," is based on a deductive procedure similar to Poe's, he has chosen for the description of his entirely imaginative, even allegorical, environment what I would call an "as-if realism"; he offers such factual details of modern life that the reader, at least at the beginning of the story, believes himself to be in a realistic milieu. We seem to find ourselves with the French Foreign Legion on some remote island; and the description, by the enthusiastic officer, of the summary methods of jurisdiction practiced in the colony or of the executions by means of an elaborately devised machine produces at first an impression of factual accuracy (written in 1919, the story seems to anticipate Hitlerism). Only in the further course of the story do we realize the fantastic character of that jurisdiction and of that gruesome engine, both of which are contrived (deductively!) by the author only in order to symbolize the inevitable, if ununderstandable cruelty of any civilization.[15]

[15] The reader might find it more natural to compare the treatment of the castle motif in Poe and Kafka. And, indeed, the castle is in both somehow the embodiment of existential fear. With Poe, however, we see the castle (the mansion) dying of "its" fears, we are with him within the castle. In Kafka's *The Castle,* we are outside the castle, which seems very much alive, although the laws according to which it functions remain unknown to the protagonist, whose existential fear is motivated by his inability ever to find his place there (a symbol of the bewilderment of modern man faced with an institutionalized world which he cannot understand, but only fear).

Poe as Symbolist

by Charles Feidelson, Jr.

The diabolism of *Moby-Dick* is more an effect than a cause of
Melville's method. Pursuing the symbolic voyage to the utmost, but
realizing at the same time its ineffectuality, Ahab is ruined, and Melville discovers that he is potentially an Ahab, the devil's partisan, the
nihilist. Poe begins where Ahab leaves off. His primary aim is the
destruction of reason, and he takes pleasure in the very horror of the
task. The gentleman who comes riding up to the house of Usher is
the personification of rational convention.[1] Like all Poe's narrators,
even the most unbalanced, he would like to cling to logic and to the
common-sense material world. But he has set out on a journey which
is designed to break up all his established categories; reason is deliberately put through the mill and emerges in fragments. The story
concerns not only the fall of Usher's house—itself a symbol of the
end of rational order—but also the shock to the narrator's assump-
tions, the dissolution of his house. The writer of the "Ms. Found in
a Bottle" is a similar type. His "habits of rigid thought," "deficiency
of imagination," and "strong relish for physical philosophy" lay him
open to an extraordinary agitation as he voyages into a region of
eccentric thought and anomalous things. Moreover, despite his intense
rationalism, the abnormal appeals to something within him. At the
beginning of the story he is estranged from country and family, and
he goes voyaging out of "a kind of nervous restlessness." The secret
aim[2] of his journey, hidden from his own conscious thought, is the

Reprinted from Symbolism and American Literature *by Charles Feidelson, Jr.*
(Chicago: The University of Chicago Press, 1953), pp. 35–42, 246–52, by permission
of The University of Chicago Press. Copyright 1953 by The University of Chicago.

[1] Poe, "The Fall of the House of Usher" (*Complete Works,* ed. J. A. Harrison
[Virginia ed.; New York, 1902], III, 273–97). Until the very end of the story the
narrator is eager to dismiss all his perceptions as "fancies," "superstitions," and
"dreams" (pp. 274, 276, 290).

[2] "Ms. Found in a Bottle" (*Complete Works,* II, 1–2).

creation of a new world by the destruction of the old. Meditating his fate, he unconsciously daubs "the word DISCOVERY" upon a sail that lies on the deck.[3] But this Emersonian motive takes on an exaggerated air because of the tension within him. Irrationality is the main characteristic of his discoveries:

> A feeling, for which I have no name, has taken possession of my soul —a sensation which will admit of no analysis, to which the lessons of by-gone times are inadequate, and for which I fear futurity itself will offer me no key. To a mind constituted like my own, the latter consideration is an evil. I shall never—I know that I shall never—be satisfied with regard to the nature of my conceptions. Yet it is not wonderful that these conceptions are indefinite, since they have their origin in sources so utterly novel. A new sense—a new entity is added to my soul.[4]

The new sense of the narrator is cultivated by the immeasurably ancient, swollen ship, hovering, like the house of Usher, on the verge of annihilation; by the incomprehensible mariners, whose eyes have "an eager and uneasy meaning"; and by the chaotic sea, stretching away to ramparts of ice that look like "the walls of the universe." The horror that his old sense feels at impending extinction is balanced by "a curiosity to penetrate the mysteries of these awful regions." He comes to see that the horror and the extinction are the necessary means to the new vision: "It is evident that we are hurrying onward to some exciting knowledge—some never-to-be-imparted secret, whose attainment is destruction." The ship whirls round a gigantic amphitheater of whiteness and plunges into the vortex.[5] The narrator makes his "Discovery" through a discipline of horror, finds a new reality through the violation of the old, and attains "exciting knowledge" through the loss of his own identity.

Just as the reason of Poe's narrator is conquered not only by the situation in which he is caught but also by something within him, Poe himself was divided between extreme rationalism and extreme

---

[3] *Ibid.*, p. 10.

[4] *Ibid.*, p. 9.

[5] *Ibid.*, pp. 10–15. This symbolic use of the vortex and of whiteness is strikingly similar to the conclusion of *Moby-Dick*. With the vortex image cf. "A Descent into the Maelström" (*Complete Works*, II, 225–47), where the seafarer experiences the same mixture of curiosity and horror. With the image of whiteness cf. the end of "Narrative of A. Gordon Pym" (III, 239–42).

hostility to reason. Together with the stories that destroy the rational mind and world, he produced the tales of ratiocination. The advocate of "indefiniteness" as a poetic principle[6] was also the author of "The Philosophy of Composition," in which the poetic process is treated as a mathematical problem.[7] Poe's extreme degradation of reason resulted from the presence of both factors inside him. He was not, like Emerson and Whitman, primarily in conflict with a rationalistic society; he was at war with himself. In addition, his ability to take a purely objective view gave a new twist to their theory of poetry. While he held with them that poetry is "indirect" and "suggestive," [8] he considered the poet a craftsman,[9] deliberately constructing the vehicle of irrationality. Again, the result was extremism.

For the rest, Poe's conception of literature was basically similar to theirs and had a similar origin. The ambiguity of Poe's metaphysics, which constitute a kind of materialistic idealism, exactly corresponds to the paradox of "process." The psychophysical world projected by the transcendentalists might be called an idealistic materialism. But, instead of attempting to describe the unity of thought and things from the side of "spirit," Poe carries out the same unification in terms of matter, infinitely rarefied "until we arrive at a matter *unparticled* —without particles—indivisible—*one*." [10] His purpose is not reduction of one term to the other but reconciliation: "The matter of which I speak is, in all respects, the very 'mind' or 'spirit' of the schools . . . and is, moreover, the 'matter' of these schools at the same time." Whether phrased in idealistic or materialistic language, the paradox is the consequence of a new category, creative motion. Poe continues:

[6] Cf. "Letter to B——" (*Complete Works*, VII, xxxvii–xxxix) and "Marginalia" (XVI, 29, 137–38).

[7] *Complete Works*, XIV, 195: ". . . the work proceeded, step by step, to its completion with the precision and rigid consequence of a mathematical problem." In "Marginalia" (XVI, 170) Poe shows an awareness of the tension between his analytic and his aesthetic tendencies.

[8] For Whitman, the poet works "by curious removes, indirections," and "seldomer tells a thing than suggests . . . it." For Poe, he aims at "a suggestive indefinitiveness of meaning." Both are attempting to define the epistemological differentia of poetry, which Emerson described more fully (*Journals*, V, 189; italics in text): ". . . these complex forms allow of the utterance of his knowledge of life by *indirections* as well as in the didactic way, and can therefore express the fluxional quantities and values which the thesis or dissertation could never give."

[9] Cf. his review of Drake and Halleck (*Complete Works*, VIII, 284–85) and "Marginalia" (XVI, 98–99).

[10] "Mesmeric Revelation" (*Complete Works*, V, 245–46; italics in text).

"The unparticled matter, in motion, is thought. . . . This thought creates. All created things are but the thoughts of God." [11] As applied to poetry, creative motion becomes "the *physical power of words.*" The thought and the thing are spoken into birth in the course of that incessant modification of form which is reality. Although Poe's God is still the unmoved mover, everything else is translated into continuous activity.[12]

These ideas, which Poe sets forth only in a half-poetic style, lie behind the ambiguity of his formal literary theory. His well-known literary doctrines hover between materialism and idealism, which cut across his odd mixture of psychological and philosophical principles. According to his psychological doctrine, beauty is "not a quality . . . but an effect," [13] a state of mind produced by a certain collocation of words. According to the philosophical doctrine, a poem is the reflection of "supernal" Beauty, "of which *through* the poem . . . we attain to but brief and indeterminate glimpses." [14] Both views were disastrous in practice: the theory of effect led to crude effects; the theory of supernal beauty led to the romantic claptrap which was Poe's stock in trade.[15] But the two doctrines met in a conception which

[11] *Ibid.,* pp. 248–49.

[12] "The Power of Words" (*Complete Works,* VI, 142–44). Although Poe never mentions "transcendentalism" except with contempt, he occasionally speaks in the language of the transcendentalist: cf. "The Island of the Fay" (IV, 193–99) and "A Chapter of Suggestions" (XIV, 186). More important are the occasions when one can see him moving toward a comparable position from an opposite starting point. Thus *Eureka,* the bulk of which assumes a finite universe and a rather deistic God, concludes with a vision of infinite process—"a novel Universe swelling into existence, and then subsiding into nothingness, at every throb of the Heart Divine." And this creative heart, Poe goes on to say, *"is our own"* (XVI, 311; italics in text). Here he is carried beyond his position in "The Power of Words," where God creates "in the beginning, *only.*"

The treatment of "matter" and "spirit" in *Eureka* (esp. pp. 308–13) is complicated by Poe's failure to see that much of what he has to say about "matter" really applies to the more general concept of "substance." The principles of "attraction" and "repulsion," which according to him *"are* Matter," are equally constitutive of any substance, for these laws turn out to be simply the logical principles of inclusion and exclusion. It follows that when Poe envisages the disappearance of matter by reason of the cessation of these principles in absolute Unity, what actually remains is not, as he supposes, pure spirit but a psychophysical totality. This Unity, as Poe says, is "Nothingness . . . to all Finite Perception" and Everything to infinite perception. Compare his picture of "matter without matter" in *Eureka* with the identical matter and spirit of "Mesmeric Revelation."

[13] "The Philosophy of Composition" (*Complete Works,* XIV, 197).

[14] "The Poetic Principle" (*Complete Works,* XIV, 273–74; italics in text).

[15] In practice, he tended to reduce his "effect" to a simple, rationally definable emotion, and he identified the beauty attainable *through* the poem with conven-

was potentially a good deal more profitable. The reflection of supernal beauty is attained "by multiform combinations among the things and thoughts of Time";[16] these combinations, in various mediums, create the psychological beauty of "effect." [17] In both cases the key to poetry is the meaningful medium, which is at once a material and a spiritual reality.[18] In both cases poems are made by novel structures of the medium, "modifications of old forms—or in other words . . . *creation of new*." [19] It is this conception of a distinctly aesthetic method and

---

tionally "beautiful" materials (see the list of "poetic" elements at the end of "The Poetic Principle"). In each case he vulgarized his own theory by substituting a preconceived "effect" or "beauty" for the unpredictable outcome of poetic form. Moreover, his particular predilections were usually in the worst romantic taste. They not only vitiated his verse but tended to obscure the really valuable aspects of his theory under appeals to the "ethereal," the "sublime," and the "heavenly." The important point of "The Poetic Principle" is buried deep in references to "the desire of the moth for the star" and to the "taint of sadness" in "all the higher manifestations of true Beauty" (*Complete Works*, XIV, 273, 279).

[16] "The Poetic Principle" (*Complete Works*, XIV, 274).

[17] The effect is constructed through "combinations of event, or tone" ("Philosophy of Composition" [*Complete Works*, XIV, 194]).

[18] Cf. "The Domain of Arnheim" (*Complete Works*, VI, 180–88), where the central importance of the medium and its dual nature are dramatized by the idea of creating a poem in the physical landscape.

[19] "The Power of Words" (*Complete Works*, VI, 142; italics in text). Cf. "Marginalia" (XVI, 87–90), in which Poe discusses "points of time where the confines of the waking world blend with those of the world of dreams." Such perceptions have "*absoluteness of novelty*," and Poe argues that they are potentially within "the *power of words*" (italics in text).

Poe was much troubled by the conception of poetic creativity, since his rationalism could not accept it, while his aesthetics demanded it. On the one hand, he went out of his way to attack Coleridge's distinction between imagination and fancy, maintaining that "the fancy as nearly creates as the imagination; and neither creates in any respect" (Review of Moore's *Alciphron* [X, 61–62]). He liked to think of imagination as the mechanical combining of "atomic" elements, which already existed as "previous combinations" ("American Prose Writers. No. 2. N. P. Willis" [XII, 38]). On the other hand, he saw that imaginative collocation of elements, in distinction from the fanciful, possessed what he called a "mystic" or "ideal" force, for "there lies beneath the transparent upper current of meaning [i.e., the completely paraphrasable prose statement] an under or *suggestive* one" (Review of Moore's *Alciphron* [X, 65–66; italics in text]). He saw that "the word ποίησις itself (creation) speaks volumes upon this point" (Review of Longfellow's *Ballads and Other Poems* [XI, 74]). In his review of Drake and Halleck (VIII, 301–2) he shows that "ideality" can hardly be found in a mere "collection of natural objects," though "each individually [be] of great beauty," for it is a function of creative form. ". . . To view such natural objects as they exist, and to behold them through the medium of words, are different things." In the line "The earth is dark but the heavens are bright," the specifically poetic level of meaning, according to Poe, is created by the word "but."

In general, whenever Poe turned against his rationalistic prejudices, he turned

content that makes Poe so hostile to the poetizing of science and ethics.[20] He wants a free hand with his "multiform combinations," unhampered by determinate rational form. Like the French Symbolists who admired him, Poe takes music as the prototype of all art, because here, in the medium that is least distinguishable from its subjective or objective reference, he finds the perfect antithesis to the language and methods of reason. His emphasis on the sound of words to the detriment of their meaning is a relatively superficial point. He aims at a much more sweeping musicality: the treatment of meaningful words as though they were the autonomous notes of a musical construct, capable of being combined without regard to rational denotation.[21] And as he persists in his elaboration of the pure poem ("this poem *per se*—this poem which is a poem and nothing more—this poem written solely for the poem's sake" [22]), his mood becomes not merely indifferent to reason but actively antirational. In order to live in the reality of creative motion, the old static reality must be destroyed; the new forms can arise only through a drastic modification of the old.

The advocate of new form is Mallarmés Poe, born to invent: "Donner un sens plus pur au mots de la tribu." [23] This Poe failed in practice. Although "indefiniteness" could mean the power to transcend

---

toward the conception of a distinctively imaginative act which is at once creative and cognitive and in which the criterion of truth is a formal consistency. See "A Chapter of Suggestions" (XIV, 187), "Mellonta Tauta" (VI, 201–6), and *Eureka* (XVI, 183, 188–98).

[20] The *loci classici* are the review of Longfellow's *Ballads and Other Poems* (*Complete Works*, XI, 68–70, 84) and "The Poetic Principle" (XIV, 271–72). Cf. also "Marginalia" (XVI, 164). Poe's antididacticism and antirealism are related to his dislike of the simile, which he regarded as merely illustrative, and of allegory, which he considered extra-poetic. Both were intrusions of rational form into the poetic realm. Cf. the review of Moore's *Alciphron* (X, 68) and the review of Hawthorne's tales (XIII, 148–49).

[21] Cf. the review of Longfellow's *Ballads and Other Poems* (*Complete Works*, XI, 74–75) and "The Poetic Principle" (XIV, 274–75). In "Marginalia" (XVI, 29, 137–38) Poe denounces any attempt to give "determinateness of expression" to music, especially any effort at "imitation." Behind his constant references to music is "the Platonic . . . μουσική," which, as Poe points out, "included not merely the harmonies of tune and time, but *proportion* generally," and which "referred to the cultivation of the Taste, in contradistinction from that of the Pure Reason" ("Marginalia" [XVI, 163]). Cf. "The Colloquy of Monos and Una" (IV, 203–4 and 204n.).

[22] "The Poetic Principle" (*Complete Works*, XIV, 272). Cf. his isolation of the "art-product" as the sole legitimate object of literary criticism ("Exordium" [XI, 7]).

[23] Mallarmé, "Le Tombeau d'Edgar Poe."

the finite words of the mob, it actually produced sloppy poems. He succeeds better in his stories, which do not carry out, but portray, his aspirations. Most of the tales play upon the wonders that lie beyond the confines of reason and upon the concurrent horror of the aberration necessary to attain them. This is Baudelaire's Poe, "l'écrivain des nerfs." [24] The narrator of "The Tell-Tale Heart" grants that he is "very, very dreadfully nervous" but insists that his disease has "sharpened . . . not dulled" his senses.[25] This story, like many others, is a variation on the theme of "perverseness," which Poe defines as "a perpetual inclination, in the teeth of our best judgment, to violate that which is *Law,* merely because we understand it to be such." [26] The center of interest in these stories is not simply the emotion of horror but the irrational state of mind, terrified at itself, yet oddly prolific. In a sense, the unnatural violation of law is a natural capacity. Perverseness, though abnormal from the standpoint of reason, is "an innate and primitive principle of human action." [27]

The most general treatment of this theme is "The Fall of the House of Usher," the subject of which is aesthetic sensibility. Poe strikes a balance between the wonder and the horror of the images that assail the narrator and preoccupy Roderick. The first glimpse of the mansion "unnerves" the visitor in a manner wholly beyond analysis; try as he may to explain his melancholy as the result of "very simple natural objects" in peculiar combination, he cannot make out the formula. At the same time, his gloom has nothing in common with the romantic love of ruin: "no goading of the imagination could torture [it] into aught of the sublime." [28] As he proceeds into the house, his unnerved sensibility becomes increasingly aware of novel possibilities. "I . . . wondered to find how unfamiliar were the fancies which ordinary images were stirring up." [29]

---

[24] Baudelaire, "Edgar Poe, sa vie et ses oeuvres," *Oeuvres complètes,* VI, xxviii.

[25] "The Tell-Tale Heart" (*Complete Works,* V, 88).

[26] "The Black Cat" (*Complete Works,* V, 146).

[27] "The Imp of the Perverse" (*Complete Works,* VI, 146–47). Cf. *ibid.,* p. 149: ". . . because our reason violently deters us from the brink, *therefore* do we the most impetuously approach it" (italics in text). The aberrant state is described in such stories as "The Man of the Crowd" (IV, 139–40), "The Pit and the Pendulum" (V, 68–69), and "A Tale of the Ragged Mountains" (V, 167–68). Most of Poe's attempts at the comic are based on the inversion or destruction of rational order: cf. "Dr. Tarr and Prof. Fether" (VI, 53–77), "The Angel of the Odd" (VI, 103–15), and "The Man That Was Used Up" (III, 259–72).

[28] "The Fall of the House of Usher" (*Complete Works,* III, 273–74).

[29] *Ibid.,* p. 277.

Throughout the story the narrator, like the house, is "falling"; he exists in a kind of suspended motion between the perception of "simple natural objects" and the neurotic perception of an aberrant world. The house itself is described as utterly decayed in every stone, yet possessed of a "specious totality." It exists on the verge of complete disintegration and transformation, standing on the brink of the dark lake where the narrator sees its "remodelled and inverted" image. The zigzag fissure that is "barely perceptible" in its front points downward into the tarn.[30] The house is poised between objective reality and a symbolic status that can be attained only by its immersion in the reflecting water and simultaneous dissolution. Usher's belief in the "sentience" of the house is another way of stating the same idea. For generations this psychophysical quality has been growing, occasioned not only by the structure of the building ("the method of collocation" of the stones) but also by "its reduplication in the still waters of the tarn." The mansion and the lake together create a distinctive world, whose evidence, verified by the narrator, is "the gradual yet certain condensation of an atmosphere of their own about the waters and the walls." Usher and his family have long inhabited this world, which exerts a "silent, yet importunate and terrible influence" upon them.[31] Their "peculiar sensibility of temperament" is inseparable from their peculiar environment, just as the name "House of Usher" includes "both the family and the family mansion." [32] But they are victims of their environment only in the sense that they have made their own milieu. The strange atmosphere is the kind of reality created by their strange habits of thought—the product, as its psychophysical quality testifies, of a fusion between the mirror of their minds and the material world. The "host of unnatural sensations" [33] that afflicts Roderick and all his clan is synonymous with the decay of their mansion. The imminent disintegration of the house is Roderick's own moment of transition between the vestiges of reason and the "exciting knowledge whose attainment is destruction."

His terror is the measure of his adherence to the old reality; his art is the reward of the new. For behind the whole story is the con-

[30] *Ibid.*, pp. 274, 276–77.

[31] *Ibid.*, p. 286. Cf. p. 276.

[32] *Ibid.*, p. 275. Cf. Usher's poem, "The Haunted Palace" (pp. 284–86) and the narrator's reference to "eye-like windows" (p. 274).

[33] *Ibid.*, p. 280. The narrator cannot connect Roderick's "Arabesque expression" with "any idea of simple humanity," just as he cannot reduce the disquieting effect of the house to any "combinations of very simple natural objects" (pp. 279, 274).

ception of an antirational art. He produces music of "singular perversion," abstract painting of an indescribable eccentricity, and poems that are "wild fantasias." He is the artistic mind *in extremis* but profiting by its own extremity. Although the specimen verses that Poe offers are unfortunate, the allusions to unorthodox music and to "pure abstractions" in painting make his point well enough. Usher's art is the last stage of the quest for novelty which begins when all art is conceived in the image of music. Antirational in genesis, it is doubly antirational in form and terrifying in its escape from the canons of reason: "The paintings over which his elaborate fancy brooded . . . grew, touch by touch, into vaguenesses at which I shuddered the more thrillingly, because I shuddered knowing not why." [34]

The denouement comes in "a tempestuous yet sternly beautiful night . . . wildly singular in its terror and its beauty." The shifting winds, careering clouds, and the "unnatural light" appeal to Roderick Usher; his disorder is one with the external chaos, which announces the final disintegration that has been impending throughout the story.[35] The movement of "The Fall of the House of Usher" is like that of "The City in the Sea," [36] from expectation to fulfilment; and in both cases it is Death that is pending and realized. Death broods over the luxuriant art of the city, and death is implicit in the crumbling house and deranged mind of the Ushers. Poe associates death, the opposite of life, with the inverted world, the opposite of reason. In "The Masque of the Red Death" the bizarre taste of the duke is the counterpart of the pestilence which his walls cannot shut out; the "spectral image" of death takes over the masquerade.[37] In "The Assignation" the hero, who is dedicated to artistic incongruity, dies in order to achieve the last full measure of eccentricity: "Like these arabesque censers, my spirit is writhing in fire, and the delirium of this scene is fashioning me for the wilder visions of that land of real dreams whither I am now rapidly departing." [38] In "The Colloquy of Monos and Una" the

---

[34] *Ibid.,* pp. 283–84. The Usher family is noted for "a passionate devotion to the intricacies, perhaps even more than to the orthodox and easily recognisable beauties, of musical science" (p. 275). The "morbid acuteness of the senses" brought on by Roderick's disease (p. 280) carries him a step further. As he says in his poem, "Spirits moving musically / To a lute's well-tunèd law" must give way to "Vast forms that move fantastically / To a discordant melody" (p. 285). He loves as much as he hates these eccentric forms.

[35] *Ibid.,* p. 291.

[36] *Complete Works,* VII, 49.

[37] *Complete Works,* IV, 250–58.

[38] *Complete Works,* II, 116, 123–24.

process of dying is a revelation: "The senses were unusually active, although eccentrically so—assuming often each other's functions at random. The taste and the smell were inextricably confounded, and became one sentiment, abnormal and intense." [39]

Yet if death, like mental derangement, creates through disorganization, it is also the loss of personal identity.[40] Just as Usher simultaneously exploits and loathes his disease, he longs for death and fears it —longs for the state of "real dream" to which he tends and fears the annihilation which that entails. This is the meaning of the relationship between Roderick and his twin sister Madeline, between whom "sympathies of a scarcely intelligible nature had always existed." [41] They are hardly distinguishable, except that Madeline is less substantial, and they come to stand for two aspects of the same individual. Although Roderick laments her seeming death, he puts her living in the tomb and cannot bring himself to rescue her; she, who when living seemed almost dead, struggles to return to life. The issue here is what Poe learnedly refers to in "Morella" as "the *principium individuationis*—the notion of that identity *which at death is or is not lost for ever.*" [42] And the issue is decided by the loss of identity, since Roderick is unable to desire that she live, and her will is unable to survive. The two collapse together; derangement is completed by dissolution. With them their world collapses into eternal flux—amid "a long

---

[39] *Complete Works,* IV, 206. Cf. the end of the world in "The Conversation of Eiros and Charmion" (IV, 6–8), where the approaching destruction is accompanied by "a wild luxuriance of foliage, utterly unknown before," and the actual dissolution is "a species of intense flame, for whose surpassing brilliancy and all-fervid heat even the angels in the high Heaven of pure knowledge have no name."

[40] The hero of "The Assignation," the original title of which was "The Visionary," lacks any permanent character. Even in life his artistic temperament approaches the impersonality which he seeks in death (*Complete Works,* II, 115): ". . . his countenance . . . had no peculiar—it had no settled predominant expression to be fastened upon the memory. . . . Not that the spirit of each rapid passion failed, at any time, to throw its own distinct image upon the mirror of that face—but that the mirror, mirror-like, retained no vestige of the passion, when the passion had departed."

[41] "The Fall of the House of Usher" (*Complete Works,* III, 289).

[42] "Morella" (*Complete Works,* II, 29; italics in text). In this passage Poe plays on the difference between "the doctrines of *Identity* as urged by Schelling" (i.e., the identity of subject and object) and "that identity which is termed personal" and which "Mr. Locke . . . truly defines to consist in the saneness of a rational being." Poe clearly lines up reason, individuality, and life on one side; irrationalism, impersonality, and death on the other. The same theme, with various emphases, is woven into stories like "Berenice" (II, 16–26), "Ligeia" (II, 248–68), and "Eleonora" (IV, 236–44). In "The Oval Portrait" (IV, 245–49) not the thinker but his object dies in the creation of the aesthetic object.

tumultuous shouting sound like the voice of a thousand waters." The lake which formerly presented the "remodelled and inverted" image of the mansion now actually closes over "the fragments" of the house of Usher.[43]

[43] "The Fall of the House of Usher" (*Complete Works*, III, 297). In a sense, by identifying art with death, Poe is saying that art, paradoxically, ceases to exist in the degree that it attains its goal. Similarly, on a more philosophical level, he maintained in *Eureka* (XVI, 310–11) that the conversion of rational structure into absolute unity simultaneously converts the world into absolute nothingness. In "The Colloquy of Monos and Una," while the "wreck and the chaos of the usual senses" are revelatory, the reality to which they lead is nihilistic (IV, 209–12). Its mere "duration" amounts to non-entity: "For *that* which *was not*—for that which had no form—for that which had no thought—for that which had no sentience—for that which was soulless, yet of which matter formed no portion—for all this nothingness, yet for all this immortality, the grave was still a home, and the corrosive hours, co-mates" (italics in text).

"That Spectre in My Path"

by Patrick F. Quinn

In the attempt to account even approximately for the permanent strength of Poe's best work it is not his sources, in the academic sense, that matter nearly so much as do his resources. Evidence of the latter kind, internal rather than external, should prove more enlightening. This is a probability which for Baudelaire amounted to a certainty. "The characters in Poe," he writes, "or rather the Poe character, the man with hyper-acute faculties, the man with ice-water in his veins . . . this is Poe himself. And his women . . . they too are Poe." If this insight is valid it leads us not to the books Poe read but to the mind that read them. Our attention should be given to what that mind produced. Thus Baudelaire directs us to the stories themselves. In this chapter, developing a point suggested in the discussion of *Arthur Gordon Pym,* we will examine a group of stories in which the theme of the Double appears. We will try to see what significance this theme has in the work of Poe and how relevant it is to our understanding of him.

* * *

Now it strikes us at once, when we turn to the detective tales, that the theme of the *Doppelgänger* will not be an important one in this division of Poe's work. For in these stories Poe attempted to exercise the rational, puzzle-solving bent of his mind, that level of his intelligence that was, in its way, scientific, logical, and systematic. On this plane, the mind is turned outward, as it were, to the light of an objective logic. A kind of game is involved, for the purpose of which the chief player renounces his ego and submits his judgment to the

"That Spectre in My Path." From The French Face of Edgar Poe *by Patrick F. Quinn, pp. 220, 223–25, 236–46. Copyright, 1954, by Patrick F. Quinn, © 1967, by Southern Illinois University Press. Reprinted by permission of the Southern Illinois University Press.*

impersonal laws of evidence and probability. The "area" of the story is not, therefore, the shadowland of psychology, but instead the well-defined field of rational inquiry. In the detective stories we seem to be following the operations of an enlightened and delicately working intelligence, and not—as in such stories as "Berenice" and "The Black Cat"—the convolutions of a deranged and tortured mind. Dupin, in whose character Poe created the archetypal detective, solves the crime; Roderick Usher, however, is the criminal, and it is Usher who typifies the damaged soul.

Dupin *vs.* Usher: that there is this basic disjunction in Poe's work is significant of more than his conformity to a conventional social-ethical outlook. For if there is no explicit glorification of the criminal-as-hero, there is no overt moralizing either. By thus avoiding the easy moral inference, Poe revealed which of his two antithetical heroes he more nearly resembled. There can be little doubt that we find the author in Usher rather than in Dupin, an indication that, despite his pretensions to a universal sort of mind, Poe's true bent was towards the darker regions of the psyche and not towards the clear and level areas of logic. In the stories of psychological terror it is the author himself who speaks as the criminal hero; whereas in the detective stories it is Dupin or Legrand who is the real hero, and Poe as narrator takes on a quite subsidiary role. He apparently found it impossible, writing in the first person, to project himself clearly as the efficiently reasoning detective. Perhaps, for one reason, he instinctively withdrew from so severely intellectual a role; and perhaps he avoided it also because, aware that the deepest dispositions of his nature were tinged with evil, he could not imagine himself as the solver of crime. Poe could not be the detective, the hunter, for he was too radically the criminal, the prey.

* * *

In me dids't thou exist—and, in my death, see by this image, which is thine own, how utterly thou hast murdered thyself.*

If Madeline Usher could have spoken when she returned from the burial vault to confront her brother, she too might have used these words; for "The Fall of the House of Usher" is also based on the *Doppelgänger* theme.

* These words are spoken to the narrator of Poe's "William Wilson" by the second self that he has just stabbed. [Editor]

All we learn of the strange situation in this extraordinary house we learn through the eyes and the emotional responses of the narrator. He is at the same time both the author of the story and, as spectator of its events, the audience as well. We react as he does. When the final catastrophe is imminent, we share his revulsion: "From that chamber, and from that mansion, I fled aghast." But there is an additional complication: the narrator is also curiously linked to the chief character. Although he cannot explain precisely why he is terrified in the house, the narrator seems to feel some kinship with Roderick Usher. Perhaps Usher's malady is not so singular as to be exclusively his. It may sometime overwhelm his visitor. The fatal events that lie ahead for Usher may also lie ahead for him. Thus when he enters the house and sees the "carvings on the ceilings, the sombre tapestries of the walls, the ebon blackness of the floors"—these prove to be "matters to which or to such as which, I had been accustomed from my infancy"; and he "hesitated not to acknowledge how familiar all this was." Familiar because he had been a boyhood friend of Usher and had come in that way to know the interior of the Usher mansion; or because the house of his own family was decorated in a similar fashion? It is probably for the latter reason. The setting in which he now finds himself is not foreign to his own mode of life, and the force of the horror which he will increasingly experience derives from his discovery of how in such a familiar setting there should be so much gone wrong. The house is underlaid with the most baffling ambiguities; not an action nor a motive has a self-evident purpose. This much at least the visitor will learn. But even as he makes his entrance he cannot clearly account for the origin of his sensations. The mist of the unknown has drifted around all he sees, and around his own mind as well. He can make things out, still recognizable, but blurred and shifting.

This is his experience from the very outset of the story. He is unable to explain the "insufferable gloom" he feels when he first comes in sight of the house. The building itself, with its vacant, eye-like windows, the rank sedge, and the dead trees—these objects should not, in themselves, oppress him. So he reasons, and in an effort to dispel this effect he studies their reversed images in the tarn that lies before the house. But this experiment only deepens his sense of gloom and foreboding. Unable to resolve the mystery here, he enters the house, in which a more complex enigma awaits him.

The two puzzles are actually very closely linked, however. Through the narrative technique he is employing in this story, Poe aligns the reader with the consciousness of the visitor to the House of Usher. Both will participate in the experience of undefined, ambiguous, and yet very palpable evil. Usher's guest never penetrates beyond the appearances; he *lives* this experience; its significance eludes him. But the reader need not be bound by such ignorance. Poe is careful to provide details of a sort that will elucidate the mystery of the House of Usher. In other words, the opening scene of the story not only serves to establish the atmosphere of doubt and misgiving, but also to suggest the moral and psychological sources from which this atmosphere emanates. What perturbs the narrator in the appearance of the house and its grounds is that he is faced with a vision of decay. It is not the condition of death which he sees, but that of death-in-life. The house, of course, is the man, an obvious representation of Roderick Usher; and of his sister also, who in her subsequent cataleptic state is neither living nor dead.

Poe uses a strikingly paradoxical figure to describe the impression which this opening scene makes on the narrator. He calls his depression of soul a sensation comparable only to "the after-dream of the reveller upon opium—the bitter lapse into every-day life—the hideous dropping off of the veil." But the narrator of this story does not come upon the conditions of everyday life at Usher's house. Rather the reverse: he has left everyday life behind him when he enters upon a scene in which decay and death are the presiding elements. His lapse is into a dreamlike state, and a hideous veil has been let down rather than removed. However, it is only through the wrenching effect of paradox that the baffling complexities of his state of mind may be conveyed. Poe uses another device to reinforce this point. When the narrator looks into the "black and lurid tarn" and sees reflected in it the house, the sedge, and the decayed trees, he experiences a "shudder ever more thrilling than before." Why? Is it not because the unreal image, the mere reflection, seems to him more real and more threatening than the actual three-dimensional house, sedge, and trees which he has just observed before? Of the two images available, it is the shadow rather than the substance that proves to be the more terrifying. How else, except through a flat, directive statement, which would dissipate altogether the essential tone of the story, could Poe indicate that "The Fall of the House of Usher" concerns the terror of the soul, and that its visible realities are of importance only as clues to the forces concealed within that are engaged in a fatal conflict?

The reflected image of the house in the water acts also as a kind of prophetic picture. The final scene of the story, in which the waters of the tarn swallow the House of Usher, is foreshadowed in this introductory episode. The house in the water has a further and more important meaning, one which relates to the almost explicit equation that is later set up between Usher and his ancestral mansion. The narrator soon becomes deeply alarmed at the appearance and conduct of his friend, but the true horror of his case is not in evidence. It resides rather in the submerged being of Usher, which is here symbolized by the reflection of the house in the water; and it is for this reason that the narrator is terrified more by the reflected image than by the actual, physical building. Thus Poe implies that the evil influences operative in the story derived from the recesses of Usher's mind. In those depths developed the impulses that led to his undoing.

"The Fall of the House of Usher" is usually alluded to, or dismissed, as a famous "atmosphere" story, and so it is; although to say only that about it is to miss more relevant sources of its rare power. To get at those sources something more is needed than the general term *atmosphere*. To be more specific, it is an atmosphere of decay, corruption, putrescence, which pervades this story. It is this that the narrator dimly becomes aware of when he raises his eyes from the tarn to the house. He sees it enveloped in "an atmosphere which had no affinity with the air of heaven, but which had reeked up from the decayed trees, and the gray wall, and the silent tarn—a pestilent and mystic vapor, dull, sluggish, faintly discernible, and leaden-hued." This atmosphere *exists*. At the climax of the story the narrator will discern it again, unmistakably. Now, with the "hideous dropping off of the veil" he speaks of, he finds it possible to make out this aura of decay. And when Poe writes that this atmosphere had no affinity with the air of heaven, he suggests that its affinity is with hell and that the inhabitants of this house are damned.

All Poe's great strokes are accomplished by implication, by suggestion. Especially in such a story as this, in which so much is staked on translucence, ambiguity, and doubt, indirection is the essence of his method. But the details are there and they are not haphazard. As the narrator examines the building more closely, he is amazed to perceive a contrast between, on the one hand, its antiquity and the crumbling condition of the individual stones, and, on the other, "the still perfect adaptation of parts." This reminds him of the "specious totality of old woodwork which has rotted for long years in some neglected vault, with no disturbance from the breath of the external air." The transition here is from stone to wood, reminding us of those decayed trees standing on the ground before the house, the trees to which he has already three times referred, and which now begin to stand as emblematic of the family line of the Ushers. All the images in this first part of the story are images of sterility and rankness. Dry rot is the fused image, now introduced through the reference to old woodwork.

Like the house which represents him, Roderick Usher also seems to emanate an atmosphere of death. The further the narrator succeeds in penetrating into the depths of Usher's mind, the more he realizes "the futility of all attempt at cheering a mind from which darkness, as if an inherent positive quality, poured forth upon all objects of the moral and physical universe, in one unceasing radiation of gloom." In the impromptu which Usher composes, "The Haunted Palace," he is represented as a "radiant" palace, although here, of course, the radiance of health is meant. And in the painting by Usher —a prophetic painting that forecasts the subterranean vault in which the body of his sister will be placed—the phenomenon of radiance is also present: "No outlet was observed in any portion of its vast extent, and no torch or other source of artificial light was discernible; yet a flood of intense rays rolled throughout, and bathed the whole in a ghastly and inappropriate splendor." Through the repetition of this detail Poe succeeds in charging the "atmosphere" of the story. This is not for macabre décor. The effect of Poe's insistence on this detail is to lend symbolic force to the phenomenon. Usher himself is aware of it. His explanation is that the very stones of the house are sentient, and together with the decayed trees, the network of fungi on the exterior of the building, and the reduplication of all this in the stagnant waters of the tarn, the unholy atmosphere has been engendered. During the night of the storm Usher shows his friend the full extent of this miasma:

A whirlwind had apparently collected its force in our vicinity. . . . But the under surfaces of the huge masses of agitated vapor, as well as all terrestrial objects immediately around us, were glowing in the unnatural light of a faintly luminous and distinctly visible gaseous exhalation which hung about and enshrouded the mansion.

By now, the appropriateness of *enshrouded* is quite clear. The House of Usher is not just yet a house of death, but it is shortly to be one.

Its stones are sentient, half-alive; and Usher himself is correspondingly half-dead. The house is a house of almost total decay. And the decay which it contains is not only physical, but, with Usher, mental and moral decay too, of which the image of dry rot is an all but precise formulation.

For what but a psychological and moral enormity could exist as the motive for Usher's burial of his sister? On a problem of this sort Poe is silent. He provides no direct indications, and, through the device of having the story told by a narrator who barely half-understands the phenomena he encounters, Poe frees himself of any responsibility for illuminating the most complex of the enigmas in the House of Usher. It is not that his almost obsessive fascination with crime suddenly ceased at the borders of the unnatural and perverse. On the contrary, this territory was of great interest to him. But, whatever may be the reason for his reluctance, he did not *directly* explore the moral issues that this area of his work involves. In this respect above all his method is implicative, and so powerfully is this method used in the story that the unnatural atmosphere which permeates the Usher mansion becomes finally redolent of moral corruption. Usher's burial of his still living sister goes beyond sadism. His act seems to culminate a relationship that in intention if not in deed was incestuous.

It is as a result of the whole contextual quality of the story, the special kind of details that are named, and those also that are not, that the relationship of Roderick and Madeline as twins takes on a sinister significance. The story is most readily intelligible as another fable of the split personality. The fissure which ran down the façade of the house and which the visitor first noticed when he made his approach to it, is a clue not only to the instability of the mind of Roderick Usher but to this particular kind of mental disorder. Roderick and his sister are the sole survivors of the Usher line; they are the House. Their deaths are simultaneous, and soon after this takes place the physical house divides from top to bottom and collapses in the waters of the tarn. In that tarn, we remember, the narrator's view of the reflection of the house caused an impression of horror on him more profound than that occasioned by the sight of the actual building. Similarly, the shadowy Madeline proved more terrifying to him than did Roderick, whom he could see clearly and to whom he could speak. He was not even aware that Madeline was in the same room with himself and Usher until "she passed through a remote portion of the apartment, and, without having noticed my presence, disappeared. I regarded her with an utter astonishment not unmingled with dread—and yet I found it impossible to account for my feelings. A sensation of stupor oppressed me, as my eyes followed her retreating steps." The two experiences correspond in a way that suggests an analogy between Madeline and the house-in-the-water. The two, the twin sister and the reflected house, represent the dark, under side of Usher's mind, the depths which cannot plainly be "read," and which are nonetheless forcibly felt to exist. And therein also is the source of the man's fatal malady. Considered in this way, "The Fall of the House of Usher" is a reversal of the Double motif of "William Wilson." Wilson was haunted by his conscience and finally killed it, and so himself. The warfare taking place in Roderick Usher is waged by his consciousness against the evil of his unconscious.

But it is obvious that his final effort in this struggle comes too late. The corrosive effort of the evil has gone too far to be withstood now. The arrival and continued presence of his friend succeed to some extent in weakening the dark side of the conflict, for Madeline succumbs soon after Usher's friend has arrived at the house. To this degree, Usher's plea for help is successful. However, his malady is too far advanced, do what he will, to permit of a cure. He hastens to inter his sister's body in an underground vault, thus fulfilling the desire of which his painting was the prophetic, and almost surrealistic, representation. But he cannot thus rid himself of what is so integrally a part of his own nature. This is the explanation for his ambiguous solicitude regarding the body of Madeline. He does not want autopsies by the "medical men," and therefore the body is interred in secret; and it is interred within the house so that there will be no risk of ghouls' prying open the coffin—something which might happen if the coffin were placed in the "remote and exposed situation of the burial ground of the family." His decision to keep the body in the house is also formed "by consideration of the unusual character of the malady of the deceased." But this reason nullifies the other two. For if Madeline were not really dead the doctors might be able to hasten her return to life, and a rifling of the grave might also serve the purpose of a resurrection. Usher, however, must have it both ways: do away with the body, and yet not do away with it altogether. For it is an integral part of himself that he is trying to dispose of. The death of Madeline, he dimly realizes, would be his own death, as it is finally. And so, during the time of her entombment, he undergoes a significant change:

His ordinary manner had vanished. His ordinary occupations were
neglected or forgotten. He roamed from chamber to chamber with
hurried, unequal, and objectless step. The pallor of his countenance
had assumed, if possible, a more ghastly hue—but the luminousness of
his eye had utterly gone out. . . . There were times, indeed, when I
thought his unceasingly agitated mind was laboring with some oppressive secret, to divulge which he struggled for the necessary courage.

Usher's secret is at last revealed: "We have put her living in the
tomb!" But his crime involved more than an attempt, strangely mis-
managed, at murder. In bringing about the premature burial of his
sister, who was his twin and counterpart, he was, like the madman
of "The Tell-Tale Heart," seeking his own death. His condition is
accordingly like that of his sister, as she lies in the vault, midway
between life and death. She returns from her grave only to take him
with her and to complete the total extinction of the House of Usher.
When Baudelaire described the "perversity" so often encountered in
Poe's stories as the quality by which a man may be simultaneously
and always the slayer and the slain, the victim and the executioner,
he defined exactly the character of Poe's most famous hero.

The Tale as Allegory

by Edward H. Davidson

One of the primary marks of a writer whose imagination might be regarded as religious but whose temper has long removed him from any doctrinal or dogmatic religious content is that, whether poet or tale-teller, he becomes his own god, his own supreme maker of visions, prophecies, and parables. Yet all the while the baffling character of these projections is that they have no apparent relationship to any body of truth or revelation. They are, to put it another way, not anti- or pro-Christian; they are simply not Christian nor even pagan; they have, if such things can be, the character of being a wholly invented simulacrum of a religious action and faith. In this respect, to play god was one of the favorite excursions of the romantic mind: Shelley engaged in the adventure so far that he invented a universe of Idea which, in an instant, he could destroy; and Keats's private pleasure-dome of aesthetic dimension had all the requisites of a profound religious experience, while Keats himself was his own god and demon. In the end, the religious mind of the Romantics became demonic because it was ultimately destructive of what it had created.

Poe's assumption of the role of god took a form not quite typical of poets or imaginative seers in the nineteenth century but one characteristic of the spell-binders and projectors of new thought in that age. His role takes him into a religious primitivism, that is, back to the primary revelation or to the original moment when the revelation was given, just as Protestant enthusiasts have longed for a return to Apostolic times and to a re-creation of the true gospel as it was initially revealed by the Messiah. In another way, however, Poe is typical of certain expressions of the Romantic mind: his religious premise is

Reprinted by permission of the publishers from Edward H. Davidson, Poe: A Critical Study (Cambridge, Mass.: The Belknap Press of Harvard University Press, 1957), pp. 186–89, 192–98, 281–82. Copyright 1957 by the President and Fellows of Harvard College.

essentially anti-intellectual and anti-ritualistic; he would return to the pure religion before it became contaminated by priestcraft and bell-ringing. Or, to state a corollary, he looks forward to the final Apocalypse, to the utter destruction of all things wherein the god finally achieves his justice or gets his awful revenge on the wicked.

In such an early tale as "The Conversation of Eiros and Charmion" (1839) we have this vision of the last day. A comet, in accordance with all the words of "the biblical prophecies," came within range of the earth and, by "a total extraction of the nitrogen" around the earth, rendered the air so combustible that the world was destroyed in one massive, blinding flash—"the entire fulfilment," Poe as vision-maker insists, "in all their minute and terrible details, of the fiery and horror-inspiring denunciations of the prophecies of the Holy Book." [1] We are not too far from the Puritan Wigglesworth's *Day of Doom* or the horrendous threats hurled against the damned by nineteenth-century evangelists.

Yet Poe's apocalyptic visions were not intended as denunciations; they were meant to be rationalizations or scientific expositions of what might be considered proved religious fact. In "The Colloquy of Monos and Una" (1841) the idea turns on not the end of the world as an inevitable fact in the logistics of nature but on the death of a single human being as a "swoon" or transfer from one form of perception to another. Monos, or the fractured and many-sided human being, passes through the three phases of being which, as we shall see, marked the upward progress toward personal fulfillment in Poe's hierarchy of insight: Monos proceeds through the physical or sensual, then through the intellectual ("a mental pendulous sensation"), and finally into pure spiritual being from which, the body having been resolved to dust, the mind and imagination can move, beyond "Place and Time." [2]

Poe's religious inquiry began, therefore, with simplicities. He took creation either back to its primal origin or forward to its ultimate consummation. Like religious myth-makers of long ago, he felt free to create his own cosmos in any form that suited him and to give it any function necessary to its fulfillment. Thus death was denied in the mere "swoon" from one stage of perception to the next, as in "The Colloquy of Monos and Una"; or he was privileged to evolve a universe in which all its atomic structures served only those purposes and

---

[1] *Works,* IV, 5, 8. Mr. Allen Tate has considered at length these apocalyptical tales; see "The Angelic Imagination: Poe and the Power of Words," *Kenyon Rev.,* XIV (Summer 1952), 455–475.
[2] See *Works,* IV, 200–212.

intentions which he as god-player ordained. It was, in short, a child's magic world, and hardly religious at all; for it had no room for evil and no condition of tragedy; no souls were lost and none was saved. But it was at such an utter remove from the conventionalized religious world in which Poe moved that it seemed like a revenge on what, in its own terms, was a rigid and institutionalized body of thought and belief. Poe's universe in imagination was at least spaceless and timeless in contrast to the easy temporality of the church in Richmond, or Philadelphia, or New York.

This universe of Poe's is not merely a spaceless and timeless cosmos; people do inhabit it; yet they do not exist in it on the simple level of morality and belief that one might expect in such a primal world as Poe imagined. Poe's nightmare universe is one in which the world is itself either just begun or just finished but the people in it are condemned to live as if they are in some long after-time of belief and morality. They exist very like the South Sea islanders Melville found on Nukaheva: they live by a rigid code of the taboo, but they have long lost any notion of what the code means. They are forced to believe and exist for reasons that have long ceased to have any meaning. No one understands or can interpret, in this moral region of Poe's lost souls, why he must be punished; yet the penalty for any moral infraction is frightful and all the more terrifying because no one had enforced it and no one knows why it must be administered. The punishment comes not from a church, a law, or even from society: it comes from some inner compulsion of the evil-doer himself who suffers from what Poe otherwise terms "perversity": he must do evil, and yet he wants to be punished and to suffer. Thus he has willed his crime, and he wills his retribution.

*   *   *

If there is a moral system in these stories, it is nebulous indeed. Poe consistently attacked the Utilitarians of his day, with their idea of "happiness" and "the greatest good to the greatest number." [3] Yet, while he could not locate the moral sense in mankind itself, he was unwilling to make the individual responsible. Even more interestingly, he did not consider that the universe itself was God's primary mistake or the outward manifestation of a cosmic tragedy; for him "the in-

---

[3] The most pointed of Poe's attacks on the Utilitarians are in the tale, "Mellonta Tauta," 1849 (see *Works*, VI, 201–205) and in *Eureka*, *Works*, XVI, 188–195; see chapter VIII.

visible spheres" were not "formed in fright." Evil or good is each man's
right and his willing; each one saves or damns himself. But the ultimate reason why man chooses or wills one or the other is far beyond
anyone's knowing; the sinner is compulsively driven by some motive
to be malignant, by some maggot in the brain which he cannot anticipate or understand but the penalty of which he is more than
willing to suffer. This need to do evil Poe placed in the idea of "perversity," man's tendency to act "for the reason that he should not."
The "assurance of the wrong or error of any action," Poe continued,
"is often the one unconquerable *force* which impels us, and alone
impels us to its prosecution. Nor will the overwhelming tendency to
do wrong for wrong's sake admit of analysis, or resolution into ulterior
elements. It is a radical, a primitive impulse—elementary." Here then
was the rationale for man's moral system and the answer to his bewildering actions which, in so many crimes, went diametrically against
any Utilitarian theory of man's willing and seeking his own happiness
or the greatest happiness to the greatest number.

Poe was content to lodge this faculty in man alone and apparently
leave him a moral freak in the world of mind and God. Yet this
"principle, the antagonist of bliss," is, however, similarly found in the
universe itself: what man is, as a fractured and disjointed being, is
but a miniature of imperfection and dispersion in the cosmic order.
By a curious variation on the myth of Adam and Eve, Poe demon-
strated that man had willed his own degradation; the earth had suffered the fatal flaw, and throughout the rest of the world's time-span
this condition of evil and suffering steadily worsens. Owing to this
defect in man and in the universe, "the world will never see . . .
that full extent of triumphant execution, in the richer domains of
art, of which the human nature is absolutely capable." Evil and suffering have become, therefore, the capacity and measure of man to feel
and know: moral sensitivity is not an act or even a thought but the
knowledge all the while that pain is the basis for life and death is
the only release from this grotesque condition of "perversity," or
man's determination to hurt and destroy himself. Thus the Poe protagonists so eagerly will their own deaths; they must plunge into "the
common vortex of unhappiness which yawns for those of pre-eminent
endowments"; only in death can they find release and peace.

Each character in Poe's moral inquiries is his own moral arbiter,
lodged in a total moral anarchy. Society has invented law and justice,
but these are mere illusion and exact no true penalty. The Poe hero

or villain is never in revolt against them, as the Romantic hero so frequently is; the Poe hero acts as if the laws of society had never even existed. The moral drama is, however, all the more terrifying because it has no rules and no reason for bringing about the end that eventually comes. It is not even a comfortably deterministic moral scheme in which whatever happens must happen; it is a moral world of an inscrutable calculus in which any one of an infinite number of results might occur.

This, then, so far as it can be sketched with any consistency, is Poe's moral cosmology, a universe of such individualism that virtually every atom has its own right and rule to exist. Within it is, of course, lodged man; but man is himself, in an almost Shakespearean way, a mirror of the universe, or the universe is a macrocosmic extension of man. The universal metaphysic is tripartite: body, mind, and soul. Every element and form in the cosmos, as Poe had suggested in "Al Aaraaf," is constituted so that it has three separate organisms and functions at the same time that these three parts are intricately interrelated to form the one and the many, the "monos" and the "una" of a universal design. Poe's final exploration of this subject was *Eureka,* written in the last years of his life.

However inexact Poe was in his outline of the tripartite organization of the material universe, he was quite explicit concerning man: man is a being formed of three separate and yet interacting forms, body, mind, and spirit. The transitions between them are so slight as sometimes to be almost indistinguishable; they form the one total "machine" that is the complete human being; they also constitute absolutely distinct functions and even parts of the human organism, and one may become, as we shall see, hypertrophied or atrophied at the expense of the other. The sources and bases for these ideas have been sufficiently well explored that we need not consider them here; suffice to say, this scheme of the human being was derived from the popular psychology of Andrew Combe and Spurzheim early in the nineteenth century. Poe was content to assume that all of his readers were so well aware of it that he did not need to go into explanatory detail; several of his noteworthy moral investigations of men are, however, built around this psychology of the tripartite organization and functioning of man.[4]

[4] For this popular version of the "science of mind" or psychology in Poe's day, see Edward Hungerford, "Poe and Phrenology," *Amer. Lit.,* II (November 1930), 209–231.

The normal, healthy human being is one in whom these three faculties are in balance; none dominates the other. But in the mysterious and chaotic condition of the universe, which is itself a duplicate of man's state of being, anything at any moment may occur in order to tip the human psyche either way, into sanity or into madness. And like the universe, the human organism, so delicately is it made and so intricately adjusted are its parts, can be turned in an instant into any one of an infinite possible conditions or states. The mind itself, the second or midway faculty of reason and direction, has no power to control either its own condition or the responses of the body and the soul; its only capacity is to speculate on whatever state of being it finds itself in at a particular moment. Neither the body nor the soul has this power: the body functions only as brute, insensitive existence; the soul, with only rare moments of perception, has the power of penetrating far beyond the limits of this sensual existence; chiefly the soul sleeps or is moribund.

In "The Fall of the House of Usher" (1839) we have an early exposition, and one of the best, of this psychic drama, a summary of Poe's ideas and method of investigating the self in disintegration.[5] The story was a study of the tripartite division and identity of the

[5] Two months before Poe's "Usher" was printed in *Burton's Gentleman's Magazine* there appeared a short article entitled "An Opinion on Dreams"; though unsigned, it was from the hand of a certain Horace Binney Wallace, known to Poe only as "William Landor." Wallace argued that the reason nothing had hitherto been known about dreams was that there had never been a correct distinction made between Mind and Soul. "I believe," Wallace went on, "man to be in himself a *Trinity,* viz. *Mind, Body,* and *Soul.*" Then he made a distinction between "dreams," which are of the mind and "proceed partly from the supernatural, and partly from natural causes," and "visions," which are of the soul and are "immaterial . . . alone." "Thus *three* portions of the *one* man seem to be most essentially different, in this way; that the body often sleeps, the mind occasionally, the soul never." The mind is situated, therefore, at center mediating between the two opposites, body and soul. The soul is continually reporting its "visions" to the mind which, in turn, though it remembers only a small fraction of them after sleeping, is still the only link between those two opposing faculties, the body and the soul. If the mind were not forgetful and so much subject to the sleepy control of the body, then we should all be aware of the illuminations and revelations which have come so vividly to saints and mystics—and sometimes to writers and poets. [H. B. Wallace], "An Opinion on Dreams," *Burton's,* V (August 1839), 105. This identification was first made by T. O. Mabbott, "Poe's Vaults," *N & Q,* no. 198 (December 1953), 542–543. The critical writing on "The Fall of the House of Usher" has always been stimulating; in recent years it has become almost an arena for critical warfare. Of the many pieces in this debate, the following are perhaps the most stimulating: Darrel Abel, "A Key to the House of Usher," *Univ. of Toronto Rev.,* XVIII (January 1949), 176–185; and D. H. Lawrence, *Studies in Classic American Literature* (New York, 1923), pp. 110–116.

self. It was, to go even further, an attempted demonstration of the theory that spirit is extended through and animating all matter, a theory confirmed by the books which Poe, and Usher, had read: Swedenborg's *Heaven and Hell,* Campanella's *City of the Sun,* and Robert Flud's *Chiromancy,* to name only a few listed in the narrative, all of which consider the material world as manifestation of the spiritual. From the opening sentence of the story we have the point-for-point identification of the external world with the human constitution. The House is the total human being, its three parts functioning as one; the outside construction of the house is like the body; the dark tarn is a mirror or the mind which can "image . . . a strange fancy," almost "a dream." The "barely perceptible fissure" which extended "from the roof of the building . . . until it became lost in the sullen waters of the tarn" is the fatal dislocation or fracture which, as the story develops, destroys the whole psychic being of which the house is the outward manifestation.

Turning now from the material to the human realms, we find that the tripartite division of the faculties is even more clearly evidenced. Usher represents the mind or intellectual aspect of the total being:

> . . . the character of his face had been at all times remarkable. A cadaverousness of complexion; an eye large, liquid, and luminous beyond comparison; lips somewhat thin and very pallid but of a surpassingly beautiful curve; . . . a finely moulded chin, speaking, in its want of prominence, of a want of moral energy.

Madeline is the sensual or physical side of this psyche: they are identical twins (Poe ignores, and so may we, the fact that identical twins cannot be of differing sex); her name is derived from Saint Mary Magdala, which means "tower"; therefore she is the lady of the house.[6]

The tale is a study of the total disintegration of a complex human being, not in any one of the three aspects of body, mind, and soul, but in all three together. Roderick Usher suffers from the diseased mind which has too long abstracted and absented itself from physical reality; in fact, the physical world, and even the physical side of himself, fills him with such repugnance that he can maintain his unique world or self of the mind only by destroying his twin sister or the physical side of himself. Madeline sickens from some mortal disease

---

[6] Further suggestive relationships between the House and the inhabitants thereof are ably treated by Maurice Beebe, "The Fall of the House of Pyncheon," *Nineteenth-Century Fiction,* XI (July 1956), 4–6.

and, when she is presumed dead, is buried in the subterranean family vaults or in a place as far remote as possible from the place of aesthetic delight wherein the mind of Roderick lives. Yet Madeline is not dead; she returns from the coffin and in one convulsive motion brings her brother to his death: the body and the mind thus die together. Very shortly afterward the House collapses, for it has all the while represented the total being of this complex body-mind relation which Poe had studied in the symbolic guise of a brother and sister relationship: "and the deep and dark tarn . . . closed sullenly and silently over the fragments of the 'House of Usher.' "

One of the curiosities of Poe's tale is that, while we have a study not only of the interrelationship of mind and body in the psychic life of a human being but also of the rapid disintegration of that being when one aspect of the self becomes hypertrophied, we have a narrative of presumed psychological inquiry with everything presented, as it were, "outside." We know no more of Roderick or of Madeline, or of the narrator for that matter, at the end than we knew at the beginning. The method is entirely pictorial, as though external objects and the configuration of the intricate material world could themselves assume a psychic dimension: not only is the material world an outward demonstration of some inner and cosmic drama but it is at every moment exhibiting that drama more strikingly than can the human actors. The two realms, material and immaterial, coexist in such exquisite balance that one can be read as a precise synecdoche of the other. The convulsive aspect of Poe's writing becomes nowhere better apparent than in his method of making the physical world of nature experience the drama more intensely than can any human being.

The Vampire Motif in
"The Fall of the House of Usher"

by Lyle H. Kendall, Jr.

The often expressed conventional interpretation of my subject is summarized and expatiated upon in Arthur Robinson's "Order and Sentience in 'The Fall of the House of Usher,'" *PMLA*, 76 (1961), 68–81. My own view of the story, although admittedly whimsical, is that in concentrating upon symbolism, upon psychological aberration, upon its connection with *Eureka* (first published some years after the story) and with certain aspects of nineteenth-century culture, critics of "The Fall of the House of Usher" have almost universally failed to recognize that it is a Gothic tale, like "Ligeia," and that a completely satisfactory and internally directed interpretation depends on vampirism, the hereditary Usher curse. Madeline is a vampire—a succubus—as the family physician well knows and as her physical appearance and effect upon the narrator sufficiently demonstrate. The terrified and ineffectual Roderick, ostensibly suffering from pernicious anemia, is her final victim.

It is not my purpose here to trace sources and analogues, for example, the body of a murdered person hidden in a makeshift coffin in the haunted wing of a castle (Clara Reeve, *The Old English Baron*, 1777), or the climactic and cataclysmic description of eerie, horrible sounds (the final chapter of Charles Maturin's *Melmoth the Wanderer*, 1820). Poe was sufficiently familiar with Gothic materials and techniques (effectively summarized in chapter seven of James R. Foster's *History of the Pre-Romantic Novel in England*, 1949), and both male and female vampires abounded in literature by the time he published

"The Vampire Motif in 'The Fall of the House of Usher,'" by Lyle H. Kendall, Jr. From College English, XXIV (March, 1963), 450–53. Copyright 1963 by the National Council of Teachers of English. Reprinted with the permission of the National Council of Teachers of English and Lyle H. Kendall, Jr.

his contribution to the genre in 1839. The bibliography of poetry, fiction, and drama appended to Montague Summers' *The Vampire* (1929) lists at least twenty-five separate works that Poe could have read, or known about, by the time he came to invent Roderick and Madeline. Among these are Southey's *Thalaba the Destroyer* (1801), Byron's *The Giaour* (1813), Polidori's *The Vampyre: A Tale by Lord Byron* (1819), Scribe's *Le Vampire* (1820), Keats's *Lamia* (1820), Hugo's *Han d'Islande* (1823), Mérimée's *La Guzla* (1827), Liddell's *The Vampire Bride* (1833), Gautier's *La Morte Amoreuse* (1836), and a host of German works—mostly bearing the title *Der Vampyr,* or something close to it—published in the 1820's. And although it was not published until 1847, I cannot forbear mentioning Thomas Prest's enormously popular *Varney the Vampire.*

Roderick is the central figure of the narrative, Poe seeming at first glance to devote less than passing attention to Madeline as a character. Her personality seems unrealized, for she appears only three times: toward the middle of the story she passes "through a remote portion of the apartment"; some days after her supposed death she is seen in her coffin, with "the mockery of a faint blush upon the bosom and the face, and that suspiciously lingering smile upon the lip which is so terrible in death"; in the final paragraph but one she reappears to die again, falling "heavily inward upon the person of her brother." These brief appearances are nevertheless fraught with darkly suggestive significance, enough to inspire D. H. Lawrence's impressionistic diagnosis, although he takes a wrong turn: "The exquisitely sensitive Roger, vibrating without resistance with his sister Madeline, more and more exquisitely, and gradually devouring her, sucking her life like a vampire in his anguish of extreme love. And she was asking to be sucked." [1]

Roderick, neither consumed by love nor acquiescent, faces a classic dilemma. He must put an end to Madeline—the lore dictates that he must drive a stake through her body in the grave—or suffer the eventuality of wasting away, dying, and becoming a vampire himself. As an intellectual he regards either course with growing horror and at length summons an old school friend, the narrator, whom Usher tentatively plans to confide in. From the outset the evidences of vampirism are calculated to overwhelm the narrator. Even before entering the house he feels the presence of supernatural evil. Reining in his horse

---

[1] *Studies in Classic American Literature* (New York, 1923), pp. 114–115.

to contemplate the "black and lurid tarn," he recalls Roderick's "wildly importunate" letter, speaking of *bodily* as well as mental disorder. He remembers that the Usher family has "been noted, time out of mind, for a peculiar sensibility of temperament, displaying itself, through long ages, in many works of exalted art, and manifested, of late, in repeated deeds of munificent yet unobtrusive charity" (a typically ironical Poe commentary upon charity as expiation). Before he rides over the causeway to the house, the visitor reflects further upon "the very remarkable fact, that the stem of the Usher race . . . had put forth, at no period, any enduring branch; in other words, that the entire family lay in the direct line of descent, and had always, with very trifling and very temporary variations [accounting for the twins], so lain."

Once within, the narrator wonders "to find how unfamiliar were the fancies which ordinary images were stirring up." On the staircase he meets the family physician, whose countenance wears a "mingled expression of low cunning [denoting knowledge of the Usher curse] and perplexity." He finds Roderick "terribly altered, in so brief a period," (an inconsistency: earlier the narrator says, "many years had elapsed since our last meeting") with lips "thin and very pallid," a skin of "ghastly pallor," oddly contrasting with the "miraculous lustre of the eye"; his manner is characterized by "incoherence—an inconsistency" and nervous agitation. He has, in fact, all the symptoms of pernicious anemia—extreme pallor, weakness, nervous and muscular affliction, alternating periods of activity and torpor—but it is an anemia, as Usher now makes perfectly clear, beyond the reach of mere medical treatment. He explains "what he conceived to be the nature of his malady . . . a constitutional and a family evil and one for which he despaired to find a remedy." He confesses that he is a "bounden slave" to an "anomalous species" of terror. Roderick discloses, further, that he is enchained by superstition in regard to the Usher house, and that "much of the peculiar gloom which thus afflicted him could be traced to a more natural and far more palpable origin—to the severe and long-continued illness—indeed to the evidently approaching dissolution—of a tenderly beloved sister." And the invalid reveals immediately that *tenderly beloved* is ironically intended by speaking with a "bitterness which I can never forget" of Madeline's impending death.

When Madeline herself now appears, at some little distance, the guest regards her with "an utter astonishment not unmingled with dread; . . . A sensation of stupor oppressed me [a characteristic reaction to the succubus] as my eyes followed her retreating steps." Roderick himself is quite evidently terror-stricken. Reluctant to grasp the import of the plain evidence with which he has so far been presented —not to mention the supernatural assault upon his own psyche—the narrator learns that Madeline's illness has been diagnosed as "of a partially cataleptical character," which is to say, to even the most casual student of necromancy, that she has the common ability of witches to enter at will upon a trance-like, death-like state of suspended animation. Her "settled apathy" and "gradual wasting away of the person" are to be accounted for by the corresponding condition in her victim.

Following Madeline's presumed death the friends occupy themselves with poring over old books that have a curiously significant connection with Usher's dilemma. Among them are the "Chiromancy" of Robert Flud, Jean D'Indagine, and De la Chambre (dealing with palmistry). Even more significantly, "One favorite volume was a small octavo edition of the 'Directorium Inquisitorium,' by the Dominican Eymeric de Gironne" (on exorcising witches and ferreting out other sorts of heretics). But Usher's "chief delight, however, was found in the perusal of an exceedingly rare and curious book in quarto Gothic —the manual of a forgotten church—the *Vigiliae Mortuorum secundum Chorum Ecclesiae Maguntinae.*" The "wild ritual of this work"—the *Watches of the Dead according to the Choir of the Church of Mainz*—is, of course, the "Black Mass."

These books fail to provide a text for Roderick, who decides to imprison Madeline, as he says, by "preserving her corpse for a fortnight (previously to its final interment,) in one of the numerous vaults within the walls of the building." Here the plodding narrator at last scents the truth:

> The brother had been led to his resolution . . . by consideration of the *unusual character of the malady of the deceased, of certain obtrusive and eager inquiries on the part of her medical men, and of the remote and exposed situation of the burial-ground of the family.* I will not deny that when I called to mind the sinister countenance of the person whom I met upon the staircase on the day of my arrival at the house, I had no desire to oppose what I regarded as at best but a harmless, and by no means an unnatural precaution (italics mine).

Alone the two friends encoffin the body and bear it to the vault. One last look at the *mocking* features of Madeline, and then the lid to the coffin is screwed down, the massive iron door secured. "Some days of bitter grief" ensue, but soon, sensing danger from a wonted quarter, Roderick Usher spends his restless hours consumed by the old horror, which he verges on confiding to the narrator: "There were times, indeed, when I thought his unceasingly agitated mind was laboring with some oppressive secret, to divulge which he struggled for the necessary courage." As he confesses later—"I *now* tell you that I heard her first feeble movement in the hollow coffin" (*hollow* in the sense that its vampiric occupant is scarcely physical in nature) —Usher is perfectly aware of Madeline's impending escape. And on the final night the guest himself suffers an experience which suggests that her evil spirit is already abroad. Endeavoring to sleep, he cannot "reason off the nervousness which had dominion over me." The room, he feels, is exerting a bewildering influence: "An irrepressible tremor gradually pervaded my frame; and, at length, there sat upon my very heart an *incubus* of utterly causeless alarm. *Shaking this off with a gasp and a struggle,* I uplifted myself upon the pillows" (italics mine), and now he hears "low and indefinite sounds." Shortly he is joined by Usher, radiating "mad hilarity" and restrained hysteria, who rushes to a casement window and throws it "freely open to the storm." It is not difficult to imagine that all the old fiendish Ushers in the distant cemetery are, disembodied, somehow present. A whirlwind (tradition-ally signalizing a spiritual presence) "had apparently collected its force in our vicinity; for there were frequent and violent alterations in the direction of the wind; and the exceeding density of the clouds . . . did not prevent our perceiving the life-like velocity with which they flew careering from all points against each other."

The last of the Ushers is persuaded to leave the window, which is closed against electrical phenomena of "ghastly origin," and the guest begins to read aloud from the "Mad Trist," whose descriptions of sound are horribly reproduced by Madeline as she leaves her prison and approaches the listeners. Roderick's final words are "a low, hur-ried, and gibbering murmur" punctuated by extraordinarily meaning-ful phrases: " 'I *dared* not speak! . . . Oh! whither shall I fly? . . . Do I not distinguish that heavy and horrible beating of her heart?' " (Again, the slow and heavy pulse is traditionally characteristic of preternatural creatures.) Poe's accentuation of the miraculous aspects of the tale continues to the end. The sister reels upon the threshold, "then, with a low moaning cry, fell heavily inward upon the person of her brother, and in her violent and now final death-agonies, bore

him to the floor a corpse, and a victim to the terrors he had anticipated." She is a vampire to the finish, and there is no escaping the
shock of absolute recognition in "From that chamber, and from that
mansion, I fled aghast."

In this view "The Fall of the House of Usher"—typical of Poe in
its exploration of abysmal degradation—creates an experience that
possesses, within itself, credibility and unity of technique once the
basic situation is granted. And from the artist's treatment of the theme,
the active existence of malignant evil in our world, emerges his partly
optimistic and partly ironic commentary: Evil in the long run feeds
incestuously upon itself, and it is self-defeating, self-consuming, self-
annihilating; the short run is another matter.

The Metamorphoses of the Circle

by Georges Poulet

In the dream as in the awakening, in stupor as in full consciousness, the mind always finds itself encircled. It is in a sphere whose walls recede or draw together, but never cease to enclose the spectator. Pleasure and terror, extreme passivity and extreme watchfulness, hyperacuity of the senses and of the intellect, are the means by which the mind recognizes the insuperable continuity of its limits. No one before Poe has shown with as much precision the essentially circumscribed character of thought. For him, that which is limitless is inconceivable. In this, he is opposed to the romantics, which in other ways he follows. He does not go beyond limits. He never stretches beyond his reach. If he likes to analyze paroxysms and frenzies, it is because they lead to an end, which is the bottom of a well or the ceiling of a tomb. If for him, man is buried alive, then man's mission is to explore the interior surfaces of his dwelling. He must learn his way about the place. It may be that this place is surrounded by death, but within it contains life. All, therefore, is not lost. In spite of its morbidity, Poe's work is saved by its intellectual power. It measures the span of the human enclosure.

Enclosure whose walls sometimes close in in the most frightful manner, but sometimes also spread out so far that they encompass the whole cosmos. Then they recede to immeasurable distances. Yet they never cease to enclose the mind:

From The Metamorphoses of the Circle *by Georges Poulet, translated by Carley Dawson and Elliott Coleman in collaboration with the author (Baltimore: the Johns Hopkins Press, 1966), Chap. XI, pp. 198–202. Copyright 1966 by the Johns Hopkins Press. Published in French as* Les Metamorphoses du Cercle. *Copyright 1961 by Librairie Plon. Reprinted by permission of the Johns Hopkins Press. Footnotes have been translated and restyled by the editor.*

Even the spiritual vision, is it not at all points arrested by the continuous golden walls of the universe?—the walls of the myriads of the
shining bodies that mere number has appeared to blend into unity?[1]

The stars therefore form a final screen for the eye; in the same way
sheer distance creates a supreme limit for the mind:

> . . . Having once passed the limits of absolutely practical admeasure-
> ment, by means of intervening objects, our ideas of distances are *one:*
> they have no variation . . . .[2]

So, too, are the cosmic spaces circumscribed; and being circumscribed, they form a whole and a sphere; or rather, an ensemble of
spheres, encased concentrically one within the other:

> I love to regard these (things) as themselves but the colossal members
> of one vast animate and sentient whole: a whole whose form (that of
> the sphere) is the most perfect and most inclusive of all . . . .[3]

> . . . We find cycle within cycle without end . . . .[4]

Endlessly, circles spread out in space. Space is not only the expanse
enveloping them, it is the milieu in which they circulate and un-
dulate. The circles, Poe says, "revolve around one far-distant center,
which is the Godhead." [5] They, and the creation which they contain,
are "a radiation from a center." [6] Taking up certain Pythagorean no-
tions, mixing with them Stoic and Platonic sources, Edgar Allan Poe
writes in *Eureka* a kind of cosmic novel which simultaneously fascinated, in their youth, both Claudel and Valéry. Here there is no infinite sphere of which the center would be everywhere and the circumference nowhere. Poe expressly rejects this Pascalian definition of
nature, or admits it only for a universe actually infinite, to which he
pays no attention. What absorbs him is the "limited sphere" [7] of the
sidereal universe. Let us suppose absolute Unity as a divisible particle

---

[1] "The Power of Words," *Works of Edgar Allan Poe,* eds. E. C. Stedman and
G. E. Woodberry, 10 vols. (Chicago, 1895), I. 236.
[2] "Marginalia," *Works,* VII. 345.
[3] "The Island of the Fay," *Works,* II. 85.
[4] *Ibid.*
[5] *Ibid.*
[6] *Eureka, Works,* IX. 46: "Absolute Unity being taken as a centre, then the exist-
ing Universe of Stars is the result of *radiation* from that centre."
[7] *Ibid.,* 63: "The Law which we call Gravity exists on account of Matter's having
been radiated, at its origin, atomically, into a limited sphere of space." And Poe

from which there is "radiated spherically—in all directions—to immeasurable but still definite distances . . . a certain inexpressibly great yet limited number of atoms." [8] At the extremity of their expansion, the atoms disposed in concentric layers, combine with each another "according to a determinate law of which the complexity, even considered by itself solely, is utterly beyond the grasp of the imagination." [9]

But if they go beyond, it is by the unimaginable complexity of these possible combinations, not by their nature, which appears strikingly similar to the working of Poe's imagination. Like it, alternatively, it diffuses and tightens: for in its immense but limited sphere the activity of the Godhead proceeds by dilations and contractions:

> Just as it is in your power to expand or to concentrate your pleasures (the absolute amount of happiness remaining always the same), so did and does a similar capability appertain to this Divine Being, who thus passes His Eternity in perpetual variation of concentrated Self and almost Infinite Self-Diffusion.[10]

Baudelaire was to say the same: "Of the vaporization and the centralization of the *Self*. All is there."

The *self* concentrated and the *self* vaporized, the *self* of the sleeper waking in the confines of his sealed chamber, and the cosmic *self*, diffused to infinity, are always the same *self* within the same circle. From the central points to the extreme there is always the same motion, varied only by such differences as are produced by the diversity of distances and the combination of elements whose place may change, but never their number. To this number nothing is added, nothing subtracted. An undulating movement covers all space, prolongs itself throughout all duration, expands or contracts the same consciousness. Space and time are a field filling thought:

> As no thought can perish, so no act is without infinite result. We moved our hands, for example, when we were dwellers of the earth, and, in so doing, we gave vibration to the atmosphere which engirdled it. This vibration was indefinitely extended, till it gave impulse to every

---

adds, in a note: "A sphere is *necessarily* limited. I prefer tautology to a chance of misconception." It seems likely that the idea of a limited cosmic sphere should have come to Poe from Kant.

[8] *Ibid.*, 46.
[9] *Ibid.*, 39.
[10] *Ibid.*, 136.

particle of the earth's air, which henceforward, *and forever,* was actuated by the one movement of the hand.[11]

Thus the same vibrating principle pervades the same span. And what is true of the universe of God is also true of the universe of the poet. As God creates the world by diffusion and combinations of His "primordial particle" in the atoms of the spheric universe, in the same way the poet diffuses and combines the elements of his spiritual life into a tale or a poem. He becomes the mental, and yet universal, milieu, where, dilating himself, contracting himself, multiplying the interaction of inner forces, a sphere is formed, which is the "sphere of action." [12] One passes from cosmology to poetic. It is a poetic of limitation:

> If, indeed, there be any one *circle* of thought distinctly and palpably marked out from amid the jarring and tumultuous chaos of human intelligence, it is that ever green and radiant Paradise which the true poet knows, and knows alone, as the *limited realm* of his authority— as the *circumscribed* Eden of his dreams.[13]

No long poem, therefore, is thoroughly beautiful because it goes beyond the inner frontiers of the mind. No long vista is constantly pleasant because the mind is lost in its vastness: "The most objectionable phase of grandeur is that of extent; the worst phase of extent, that of distance. It is at war with the sentiment and with the sense of seclusion." [14] Like the domain of Arnheim, the space of the poem should be enclosed. It should also be spherical: "It has always appeared to me that a close *circumscription of space* is absolutely necessary to the effect of insulated incident." [15] The more space is circumscribed, the more it satisfies Poe. This is why he loves *Marginalia*: "The circumscription of space, in these pencilings, has in it something more of advantage than inconvenience." [16] Surrounded, therefore, by a circumference, reduced to the exact space it occupies, the poem or the tale exists only in itself. It concentrates or dilates within the interior surface of its sphere; it is significant only in "the reciprocity of

---

[11] "The Power of Words," *Works,* I. 238.

[12] Review of Hawthorne's Tales, *Works,* VII, 33: "The impressions produced [by Hawthorne's tales] were wrought in a legitimate *sphere of action.* . . ."

[13] Review of "The Culprit Fay," by Joseph Rodman Drake, and "Alnwick Castle," by Fitz Greene Halleck, *Complete Works of Edgar Allan Poe,* ed. J. A. Harrison (New York, 1902), VIII, 281.

[14] "The Domain of Arnheim," *Works,* eds. Stedman and Woodberry, II. 105.

[15] "The Philosophy of Composition," *Works,* VI. 42.

[16] "Marginalia," *Works,* VII. 210.

its causes and effects." [17] In the poem every element depends upon every other. Nothing is lost, nothing is added. No atom is displaced without causing everything to be affected. "Let the poet press his finger steadily upon each key, keeping it down and imagine each prolonged series of undulations." [18] "[The plot] is that from which no component atom can be removed, and in which none of the component atoms can be displaced, without ruin to the whole." [19]

No better application of this statement can be found than in *The Fall of the House of Usher*:

> I had so worked upon my imagination as really to believe that about the whole mansion and domain there hung an atmosphere peculiar to themselves and their immediate vicinity—an atmosphere which had no affinity with the air of heaven, but which had reeked up from the decayed trees, and the grey wall, and the silent tarn . . . .[20]

This atmosphere is a sphere. Without affinity to the air of heaven, reflected in the waters of its own tarn, the House of Usher exists only in the dense vapor issued from its ground. It has, so to speak, created its own space. It has also created its own particular duration. Not only does it exist in the spherical continuity of its own surroundings, but also in the linear continuity of the family it shelters. This "has perpetuated itself in direct lineage." So to the absence of connection with the air of heaven there must be added the "absence of the collateral branch." In the same way in which the house is enclosed in its own singular atmosphere, so too, its inhabitants are the prisoners of their own time, which cannot be mingled with that of the outside world. The temporal circumscription is no less rigorous than the spatial one. The two are united to "mould the destinies." "Man," says Leo Spitzer, "is now embedded in a milieu which may enclose him protectively, but may also represent his doom." [21] This doom is brought about, not by an external crisis, but by the condensation of a whirlwind whose origin stems from the particular atmosphere. When sinking into its own pool, the House of Usher disappears into itself. It reabsorbs its space and its duration. It completes, from cause to

[17] "The American Drama," *Works*, VI. 212: "The pleasure which we derive from any exertion of human ingenuity is in direct ratio of the approach of this species of reciprocity between cause and effect."
[18] "A Chapter of Suggestions," *Works*, VIII. 331.
[19] *Ibid.*, 329.
[20] "The Fall of the House of Usher," *Works*, I. 134.
[21] Leo Spitzer, "A Reinterpretation of 'The Fall of the House of Usher,'" *Comparative Literature*, IV (1952), 360.

effects, the closed cycle of its existence and of the existence of its hero.

We have therefore here an exact symbol of the "totality of effect" which is the aim of Poe's art. As in the compact volume of a sphere, everything is in rapport with everything else, and with nothing but this all. And once again, as in the sphere, everything comes back to a central point.

Edgar Poe: Style as Pose

by James M. Cox

There should no longer be any question—indeed, there probably never should have been a question—that Poe is one of our major writers. Yet in the august company of Hawthorne, Melville, Emerson, Thoreau, and Whitman, he alone is likely to have his credentials repeatedly challenged, as if he might actually be an impostor. Whatever their deficiencies as writers, his great contemporaries inescapably possess the bearing of serious artists. Poe, however, although he grandiosely proclaimed a theory of pure art, betrays an air of pretentiousness, posturing, and even downright fraud. To be sure, he has his devoted followers who see him as he wished to be seen: the embodiment of the Romantic Artist as Victim. And he has the sturdy corps of academic specialists and defenders seeking to protect his honor and reputation. Finally, he has more than his share of psychoanalytically minded critics seeking to define the nature of his threatened ego.

For Poe's life cast him in the rôle of victim—victim of orphanage, of an insensitive foster father, of alcohol, of grinding poverty, of a hostile and materialistic society, and finally of a villainous literary executor, one Rufus Griswold, whose present claim to immortality is his energetic effort to defame Poe. Small wonder that his admirers identify with his victimization, that scholars defend his sullied honor, and that psychoanalytic critics seek the primal psychic wound which bled into his art. Yet accompanying this figure of Poe is a disturbing set of contrivances which seem almost designed to provoke precisely such a response. There is in almost everything he wrote or did a certain shameless dramatization, a tawdry theatricality, which should remind posterity—if it needs reminding—that he was indeed the son

From "Edgar Poe: Style as Pose" by James M. Cox. From The Virginia Quarterly Review, XLIV (Winter, 1968), 67–71, 78, 80–81, 82, 89. Copyright 1968 by The Virginia Quarterly Review, The University of Virginia. Reprinted by permission of The Virginia Quarterly Review and James M. Cox.

of traveling actors. In other words, Poe's life constantly presents itself as if it were as much act as action, and it is difficult to escape the conclusion that at the end of his life Poe, like the diabolical narrator of "The Cask of Amontillado," deliberately trapped his hated enemy Griswold by naming him his literary executor. If so, the unsuspecting Griswold fatuously rose to the bait, producing the intensely hostile obituary which has never ceased to bring a host of scholars to Poe's defense to pronounce Griswold's distortions the act of an unprincipled scoundrel.

If there is something contrived about Poe's life, there is also something contrived about his art. This exposure of contrivance is not an error into which Poe occasionally lapses; it is an integral aspect of his identity as a writer. Aldous Huxley had the quality clearly in focus when he cited Poe as the example par excellence of vulgarity in literature. "Was Edgar Poe a major poet?" Huxley rhetorically asked himself, and confidently replied, "It would surely never occur to any English-speaking critic to say so." Despite the monolithic assurance of his English instincts, Huxley was troubled by the extreme praise fairly lavished on Poe by Baudelaire, Mallarmé, and Valéry. It was in fact the French praise in the face of Poe's patent vulgarity which struck Huxley, as it has struck many another critic of Poe, as a paradox deserving critical attention. How could this poet who thrust himself forward in the world of letters like a gentleman exhibiting a diamond ring on every finger—how could such a man, wondered Huxley, be taken seriously as a great writer? Huxley concluded that the French, while they recognized and were delighted by the refinement of Poe's substance, were at the same time blinded by virtue of the language barrier to the essential vulgarity of his form.

After thirty years Huxley's remarks still retain a singular aptness. They are as hard to explain away as that French praise which so troubled Huxley. For Poe is vulgar, if by vulgarity is meant the deliberate effort to achieve sensational effects in order to shock the sensibility of the audience. Henry James had Poe's vulgarity thoroughly in mind when he remarked that "an enthusiasm for Poe is the mark of a decidedly primitive stage of reflection." So did Paul Elmer More when he observed that "Poe is the poet of unripe boys and unsound men." So did James Russell Lowell when he found Poe three-fifths sheer genius and two-fifths sheer fudge. And so of course did Emerson when he scornfully referred to Poe as the jingle man.

Even Allen Tate, the most sympathetic and perceptive of Poe's critics in our own time, is reduced to the following admission when confronted by Poe's style:

> I confess that Poe's serious style at its typical worst makes the reading of more than one story at a sitting an almost insuperable task. The Gothic glooms, the Venetian interiors, the ancient wine cellars (from which nobody ever enjoys a vintage but always drinks "deep")—all this, done up in a glutinous prose, so fatigues one's attention that with the best will in the world one gives up, unless one gets a clue to the power underlying the flummery.

Tate speaks of Poe's style at its typical worst because he realizes just how much this worst is typical of Poe's writing. For Poe's style is so ridden with clichés that it seems always something half borrowed, half patched. And not in the worst stories only is this evident, but in the best. Here are the opening sentences of "William Wilson."

> Let me call myself, for the present, William Wilson. The fair page now lying before me need not be sullied with my real appelation. This has been already too much an object for the scorn—for the horror—for the detestation of my race. To the uttermost regions of the globe have not the indignant winds bruited its unparalleled infamy? Oh, outcast of all outcasts most abandoned!—to the earth art thou not forever dead? to its honors, to its flowers, to its golden aspirations?—and a cloud, dense, dismal, limitless, does it not hang eternally between thy hopes and heaven?

But why go on? William Wilson sounds like a fugitive from an asylum devoted expressly to the maintenance of ineffectual heroes escaped from sentimental and gothic romance.

It is of course possible to argue that this language is William Wilson's, not Poe's. The truth is, however, that all of Poe's narrators are remarkably similar—are in effect a single narrator who tells, under various names, practically all of Poe's stories. There is really no fallacy in equating this narrator's style with Poe's style so long as one does not go on to insist that Poe's narrator is Poe. For insofar as the narrator embodies Poe's narrative style he is just so much the style and not the man.

As style, the narrator is characterized by an excessive impersonation of the conventions of learning and literature which produces an effect of intellectual and literary posturing. Moreover, the narrator's literary or "narrative" posture is never separate from but invariably a part of his intellectual arrogance. These twin postures are not an accident of

Poe's style but its essence. It is not too much to say that, for Poe, style was pure pose. In a world where style is pose, there are necessarily going to be some momentous transformations. Symbol in such a world becomes anagram, form becomes rationale, imagination becomes impersonation, cause becomes effect, and creation becomes invention.

To recognize such transformations is to begin to grasp the terminology for describing the world of Edgar Poe. It is not surprising that Poe's genius, which first displayed itself in excessively rhythmic poetry, moved next to the extravagant improbabilities of sensational fiction, on to formulate a poetics of short fiction which elevated the traditional oral tale to the status of written art, before realizing itself in the invention of a new form—the detective story—which Poe characteristically and accurately termed the tale of ratiocination. For Poe had, from the very outset of his career, passionately believed that true genius was to be equated with originality, and he never ceased to celebrate the notion that in art as well as in experience the true excitement was the thrill of doing something utterly new.

*   *   *

In "Ligeia," "The Fall of the House of Usher," and "William Wilson," stories following hard upon "Pym," Poe explored the perverse world he commanded. Each of the three stories discovers its reality—which is to say its effect—in perverse disintegration of the psyche. In each story the narrator witnesses or enacts a crime which will be so thrilling in its effect as to shock him not to death but to life. Moreover, the disintegration and the crime emerge along burlesque patterns, the stories wrecking the forms upon which they prey.

*   *   *

What the story marvelously succeeds in doing is to define the relation between the two empty traditions Poe inherited and burlesqued: the gothic world of vampires and the romantic world of maidens.

"The Fall of the House of Usher" enacts the entire collapse of these traditions. When the gothic machinery of the house of Usher tumbles into the tarn at its base, it carries with it the last extremity of the romantic artist in the person of Roderick Usher. But this time the narrator, instead of assuming the pose of central actor, comes to the aid of his dying friend and in the process manages to become Usher's accomplice in burying the lady Madeline Usher alive. The closing action of the story, in which the lady Madeline claws her way out of her tomb to kill her brother, is even more insistent in its burlesque than "Ligeia." Whereas "Ligeia" extravagantly burlesqued gothic and sentimental traditions in the persons of Ligeia and Lady Rowena, Poe here goes so far as to manufacture as part of the action an overt parody of gothic romance. After the burial of the Lady Madeline, the narrator, attempting to quiet the high-strung Usher, takes down the "Mad Trist" of Sir Launcelot Canning, one of Usher's favorite romances. As he reads the impossible prose of that archaic production, the action of the narrative not only begins to conform to, but luridly exceeds the ponderous gothicism of the "Mad Trist." In this excessive impersonation of a prior degenerate form Poe literally invents a burlesque romance as a means of exposing Usher's utter degeneration. For Allen Tate, this last piece of flummery is the tastelessness which alienates the adult from the story he identified with as a child. Yet surely the ending is a full exposure of the play upon which the story is built. For Usher is, in the last analysis, the artist—the sick artist gradually dying in the stifling environment of the gothic house he haunts. Having fed upon its own decay, his imagination at last betrays itself in incest and madness. Both he and Ligeia are the decadent artists who haunt the narrator; they are his madness, his disease, and the rather sick and banal correspondence poems which each has written—and which Poe inserted grandiloquently at the center of each story—reinforce their impotence.

In "Ligeia" the maddened narrator had destroyed the blond and blue-eyed lady of romance who had repressed, distorted, and driven underground his will, which in turn took its revenge in the form of the aggressive fantasy of Ligeia. In "The Fall of the House of Usher" he brought down the House of Usher along with the pale, sensitive artist who was in turn victimized by all the degenerate traditions embodied in the house, the sister, and his art. But in "William Wilson," instead of figuring forth the drama of repression in terms of an attempt to murder a wife or bury alive a sister, the narrator directs his attack completely upon himself in an effort to kill his conscience.

*   *   *

The murder of the conscience is the crime of the Poe narrator pushed to its furthest degree. It is the crime of Pym, of the murderer of Rowena, of Usher, carried out explicitly against the self. The earlier narrators had attempted to bury their guilt in the person of the "other"; Wilson also tries, but discovers that the "other" is relentlessly himself. Having killed the other Wilson, he presumably sinks into a life of total profligacy, out of which he at last emerges to recall the story of his demise in its earliest form.

*   *   *

Poe's "art," which converted the traditional form of the tale of terror into the conscious form of the short story, reveals that terror is the illusory life into which man perpetually flees in an effort to escape the living death he cannot acknowledge. This larger terror, this perversity, is the recognition which Poe's form enacts. Since the recognition is enacted through impersonation, the faces of terror and sentiment inevitably assume lurid, grotesque, sensational, morbid, and ludicrously exaggerated postures, inviting a cultured audience to shrink from their vulgarity. But to reject Poe's art is to forego and drive underground the power of his discovery. T. S. Eliot's careful tribute to this underground current which, emerging in the poetry of Baudelaire, Mallarmé, and Valéry, flowed back into Eliot's own verse, has shown clearly, if patronizingly, how Poe made his way back to American shores in poetry.

Chronology of Important Dates

Poe	Historical and Cultural Events
1809 Born in Boston. Father disappears in 1810. Mother dies in 1811.	Washington Irving, *Knickerbocker's History of New York.*
1815 Travels with his guardians, the Allans, to Great Britain.	Napoleon defeated at Waterloo. Congress of Vienna.
1819	Irving, *Sketch-Book.*
1823	Monroe Doctrine. James Fenimore Cooper, *The Pioneers.*
1826 Enters University of Virginia, but soon withdraws because John Allan refuses to pay his gambling debts.	
1827 First volume of poems published.	Sir Walter Scott, "On the Supernatural in Fictitious Composition."
1829 *Al Aaraaf, Tamerlane, and Minor Poems.*	Andrew Jackson becomes president.
1832	British Reform Bill. Tennyson, *Poems.* Hawthorne, "My Kinsman, Major Molineux."
1833 "Ms. Found in a Bottle" wins newspaper prize.	
1835 Associated with *Southern Literary Messenger* in Richmond.	Tocqueville, *Democracy in America.*
1836 Marries his first cousin, Virginia Clemm.	Emerson, *Nature.*

1837		Hawthorne, *Twice-Told Tales*. Dickens begins serial publication of *Oliver Twist*.
1838	*The Narrative of Arthur Gordon Pym*, "Ligeia."	
1839	"The Fall of the House of Usher," *Tales of the Grotesque and Arabesque*.	Longfellow, *Voices in the Night*.
1843	Proposed magazine *The Stylus* fails. "The Gold Bug" wins prize.	Ruskin, *Modern Painters*. Kierkegaard, *Either/Or*. Carlyle, *Past and Present*.
1845	Begins editorship of *The Broadway Journal*. *The Raven and Other Poems*, revised *Tales*.	Thoreau goes to Walden Pond. Margaret Fuller, *Woman in the Nineteenth Century*.
1846		Mexican War begins. Hawthorne, *Mosses from an Old Manse*.
1847	Virginia Poe dies.	
1848	Publishes *Eureka*.	Marx and Engels, *The Communist Manifesto*.
1849	Dies suddenly in Baltimore.	California gold rush. Thoreau, "Civil Disobedience."
1850		Hawthorne, *The Scarlet Letter*.
1851		Melville, *Moby-Dick*.

Notes on the Editor and Contributors

THOMAS WOODSON, Associate Professor of English at The Ohio State University, has published articles on Melville, Thoreau, and Robert Lowell. He was Fulbright Lecturer in Pau, France, for 1968–69.

DARREL ABEL, Professor of English at Purdue University, has published a number of essays on the writers of the American Renaissance, and on American literature generally.

Princess MARIE BONAPARTE (1882–1962) was born in France, and married into the Greek royal family. During the Balkan War of 1912 she operated a hospital ship, and later wrote a book on the social effects of war. She turned to psychology, translating Freud into French, and contributing a number of articles to the *Revue francaise de psychanalyse*. Her study of Poe first appeared in 1933.

WAYNE C. BOOTH, George M. Pullman Professor of English and Dean of the College at the University of Chicago, has written widely on the techniques of fiction.

CLEANTH BROOKS, Gray Professor of Rhetoric at Yale, was Cultural Attaché to the American Embassy in London from 1964 to 1966. His *Modern Poetry and the Tradition* (1939) and *The Well-Wrought Urn* (1947) helped create a revolution in American criticism.

JAMES M. COX is the author of *Mark Twain: The Fate of Humor* and editor of the Twentieth Century Views volume on Robert Frost. He is Professor of English at Dartmouth College.

EDWARD H. DAVIDSON, Professor of English at the University of Illinois, has written books on Hawthorne, Poe, and Jonathan Edwards.

CHARLES FEIDELSON, Jr., Professor of English at Yale, recently edited, with Richard Ellmann, *The Modern Tradition*. His influential *Symbolism and American Literature* was published in 1953.

CAROLINE GORDON has been Lecturer in Creative Writing at the School of General Studies, Columbia University, since 1946. Her fiction includes *The Malefactors* (1956) and *Old Red and Other Stories* (1963).

LYLE H. KENDALL, Jr., Professor of English at the University of Texas at Arlington, is author of a critical biography of George Wither, and coauthor of *A College Rhetoric*.

D. H. LAWRENCE (1885–1930), the great English novelist and poet, published his *Studies in Classic American Literature* in 1923.

HARRY LEVIN has written on such diverse subjects as Ben Jonson, Joyce, Shakespeare, Hawthorne, Marlow, Melville, and the French realistic novel. He is Irving Babbitt Professor of Comparative Literature at Harvard.

GEORGES POULET, Professor of French at the University of Zurich, formerly taught at the University of Edinburgh and at Johns Hopkins. His books translated into English include *The Interior Distance* and *Studies in Human Time*, which contains an appendix on "Time and American Writers."

ARTHUR HOBSON QUINN (1875–1960), Professor of English at the University of Pennsylvania, was author of histories of American drama and fiction.

PATRICK F. QUINN, Professor of English at Wellesley College, has published essays on Emerson and Henry James as well as several studies of Poe.

LEO SPITZER (1887–1960) was Professor of Romance Philology at several Austrian and German universities, and, from 1936 until his death, at Johns Hopkins. His works in English include *Linguistics and Literary History* (1948) and *Classical and Christian Ideas of World Harmony* (1963).

ALLEN TATE, poet and critic, Regents' Professor of English at the University of Minnesota, has included several studies of Poe in his *Collected Essays* (1960).

ROBERT PENN WARREN, novelist, poet, and critic, is Professor of English at Yale. He collaborated with Cleanth Brooks in several influential textbooks, beginning with *Understanding Poetry* (1938). His novel *World Enough and Time* uses the same historical materials as Poe's verse tragedy *Politian*.

Selected Bibliography

Bailey, J. O., "What Happens in 'The Fall of the House of Usher?'" *American Literature*, XXXV (1964), 445–66. Speculates that Poe intentionally concealed the story's "basis of terror"—the lore and literature of vampires —by refining it to "the strange and mystical."

Beebe, Maurice, "The Universe of Roderick Usher," in *Ivory Towers and Sacred Founts*, pp. 118–28. New York: New York University Press, 1964. Reprinted in *Poe: A Collection of Critical Essays*, ed. Robert Regan. Englewood Cliffs, N.J.: Prentice-Hall, Inc., 1967. Sees Roderick as "a prototype of the artist-as-God," and the story as a symbolic statement of Poe's cosmological views, especially as he later expressed them in *Eureka*.

Cohen, Hennig, "Roderick Usher's Tragic Struggle," *Nineteenth Century Fiction*, XIV (1959), 270–72. Another defense against Brooks and Warren: Roderick's struggle is between his will to live and the price he must pay in order to stay alive.

Goodwin, K. L., "Roderick Usher's Overrated Knowledge," *Nineteenth Century Fiction*, XVI (1961), 173–75. Argues against Cohen that Roderick's own words indicate that he does not struggle to live because of his strong bond to Madeline.

Hoffman, Michael J., "The House of Usher and Negative Romanticism," *Studies in Romanticism*, IV (1965), 158–68. Sees the House as a symbol of the Enlightenment, Roderick as prisoner of the old rationalism, and the narrator as a nameless new post-Enlightenment man, a "Negative Romantic."

Olson, Bruce, "Poe's Strategy in 'The Fall of the House of Usher,'" *Modern Language Notes*, LXXV (1960), 556–59. The story itself demonstrates the superiority of the artistic imagination to the analytical intellect.

Robinson, E. Arthur, "Order and Sentience in 'The Fall of the House of Usher,'" *PMLA*, LXXVI (1961), 68–81. The principle of order is the key to a rigorously rational pattern of thought and thematic development behind the mood and characterization of the story.

Smith, Herbert F., "Usher's Madness and Poe's Organicism: A Source," *American Literature*, XXXIX (1967), 379–89. Compares Roderick Usher on "the sentience of all vegetable things" to Richard Watson's *Chemical Essay* (mentioned by Poe in a footnote).

Walker, I. M., "The 'Legitimate Sources' of Terror in 'The Fall of the House of Usher,' " *Modern Language Review*, LXI (1966), 585–92. Contemporary science explains Roderick's disintegration as the effect of inhaling noxious vapors from the tarn.

Wilbur, Richard, "Introduction," to *Poe: Complete Poems*, pp. 19–28. New York: Dell Publishing Company, 1959. (See also Wilbur's "The House of Poe," in Regan, *Poe: A Collection of Critical Essays*.) "Usher" as a "dream-voyage" like "Ms. Found in a Bottle" and "The Domain of Arnheim"; the narrator and Roderick are complementary aspects of the hero's "Bi-Part Soul": Poet and Dreamer.

TWENTIETH CENTURY
INTERPRETATIONS

MAYNARD MACK, *Series Editor*
Yale University

NOW AVAILABLE
Collections of Critical Essays
ON

ADVENTURES OF HUCKLEBERRY FINN
ALL FOR LOVE
THE AMBASSADORS
ARROWSMITH
AS YOU LIKE IT
BLEAK HOUSE
THE BOOK OF JOB
THE CASTLE
DOCTOR FAUSTUS
DUBLINERS
THE DUCHESS OF MALFI
EURIPIDES' ALCESTIS
THE FALL OF THE HOUSE OF USHER
THE FROGS
GRAY'S ELEGY
THE GREAT GATSBY
GULLIVER'S TRAVELS
HAMLET
HARD TIMES
HENRY IV, PART TWO
HENRY V
THE ICEMAN COMETH
JULIUS CAESAR

(continued on next page)

(*continued from previous page*)

KEATS'S ODES
LORD JIM
MUCH ADO ABOUT NOTHING
OEDIPUS REX
THE OLD MAN AND THE SEA
PAMELA
THE PLAYBOY OF THE WESTERN WORLD
THE PORTRAIT OF A LADY
A PORTRAIT OF THE ARTIST AS A YOUNG MAN
PRIDE AND PREJUDICE
THE RAPE OF THE LOCK
THE RIME OF THE ANCIENT MARINER
ROBINSON CRUSOE
SAMSON AGONISTES
THE SCARLET LETTER
SIR GAWAIN AND THE GREEN KNIGHT
THE SOUND AND THE FURY
THE TEMPEST
TOM JONES
TWELFTH NIGHT
UTOPIA
WALDEN
THE WASTE LAND
WUTHERING HEIGHTS